Independent Generation of Electric Power

Independent Generation of Electric Power

David Stephen

Butterworth-Heinemann Ltd
Linacre House, Jordan Hill, Oxford OX2 8DP

A member of the Reed Elsevier Group

OXFORD LONDON BOSTON
MUNICH NEW DELHI SINGAPORE SYDNEY
TOKYO TORONTO WELLINGTON

First published 1994

British Library Cataloguing in Publication Data
Stephen, David
Independent Generation of Electric Power
I. Title
621.31

ISBN 0 7506 1691 1

Library of Congress Cataloguing in Publication Data
Stephen, David.
Independent generation of electric power/David Stephen.
p. cm.
Includes index.
ISBN 0 7506 1691 1
1. Remote area power supply systems. I. Title.
TK1005.S6958 93–46164
621.31'21–dc20 CIP

Composition by Genesis Typesetting, Rochester, Kent

Printed and Bound in Great Britain by
Hartnolls Limited, Bodmin, Cornwall.

Contents

Preface

Although the basic principles of electric power generation have been known and applied for many years now, suitable control procedures have developed more slowly, dependent largely on the associated technology currently available. As a consequence the design of generation units tended to become stereotyped, and assumptions made were accepted as valid without any reevaluation being made of whether the original basis on which they had been made was still correct.

The tremendous impact of the micro-chip, and its many ramifications, on all aspects of data handling and plant control has produced a sudden and massive change in this situation, but, because of the multiplicity of engineering disciplines involved and the rapid development of new technologies, maximum benefit has not yet been obtained. Until a suitable philosophy of design including all these new factors has been fully developed and is available for general application, full advantage cannot be derived and optimum utilization of generation plants will not be realized.

This is of the greatest significance to engineers in the basic design of generation plant, either for new installations or as amendments to an existing system. They cannot hope to have expert knowledge in all the most up-to-date technologies available necessary to ensure that the most suitable plant is selected, installed and utilized correctly, in conjunction with any existing equipment. They must therefore make use of the knowledge and technology already available to engineers experienced in the practical application of the new philosophy. To do this they must obviously be aware of all the factors relevant to the design of a suitable system even if they do not themselves employ the relevant technologies.

The simplest form of generation system is the independent generation unit, of which the number of installations is rapidly increasing due to a variety of causes. These include the growing requirements for developing countries, the need for electrical power in isolated locations using high power loadings, e.g. ships, offshore platforms and remote industrial plants and processes, and the economic pressure to include CHP installations on small or specialized process plants to reduce costs and increase their reliability. The privatization of the electrical industry in the UK is also a contributing factor.

Many such systems have to be capable of operation in connection with a larger generation system and this involves a different set of requirements: the independent system has to be designed to operate satisfactorily under either regime.

It is no longer satisfactory to select standard conventional components for a generation system and expect that they will operate together to give a level of system performance acceptable to the power users. To design an independent generation system requires a fundamental assessment of all aspects of generation, distribution and utilization, and this involves the application of several engineering disciplines in considerable depth. It is not always possible to obtain the necessary depth of knowledge in all the relevant technologies involved from the staff who may be responsible for the design, installation and operation of the plant, but if they are aware of the scope and nature of the problems involved and of how these can be tackled they will be able to perform their function satisfactorily, given the necessary specialist assistance where necessary. They are left with a full understanding of the requirements of the system and the procedures adopted to ensure its satisfactory operation.

This book establishes clearly what information is required and how it can be obtained and used to select equipment and operational procedures to meet the required operating criteria. It identifies alternatives available and indicates the relative merits of these. From the information available it is possible to determine what compromises will be most beneficial for any particular set of conditions. The first chapters cover all aspects of the performance of electrical generation sets as individual units, consider the significance of alternative combinations of these, and finally explain the effects of interconnecting sets through impedances such as distribution lines and transformers. The control procedures suitable for all these arrangements are analysed, together with the special problems associated with the design of fully automated systems for simple power management and also for the more complex energy management.

For the benefit of engineers who wish to pursue particular aspects of the problems discussed in greater detail, the appendices include information on subjects commonly involved in generation projects and which are not generally available in a simple but comprehensive explanation which can be readily understood and appreciated by the non-specialist engineer.

It will be appreciated from the information provided that while it is relatively easy to adopt a solution to any generation design problem, this may not be the most desirable from any point of view, whether it be simplicity of equipment, cheapest installation, most efficient or most reliable plant, and may, in fact, not even be capable of performing its function to the satisfaction of the user under actual operating conditions. It is necessary for the project or design engineer to evaluate the installation and operational requirements and assess all the factors described in the book, selecting one or more possible solutions which meet most closely the limiting criteria, and from these decide the overall optimum for the particular installation.

Introduction

The earliest electrical power generation units consisted of a single generator supplying a specific local load, such as domestic lighting. Later, several generating sets would be installed in a local generating station and operated in parallel to supply any electrical load in the near vicinity. Some such stations were set up by local companies for their own use or to sell power to other consumers. Many local corporations operated their own independent units as a public service to the community. However, the reliability and performance of such stations was very variable and it was found that a better and cheaper supply could be provided by generating stations that incorporated larger and more efficient units; by providing a suitable power distribution network to cover a wide area, a larger prospective electrical load could be obtained. This development continued and resulted in the UK in a single, fully interconnected electricity generation and distribution system which was required to produce a public supply to limits of performance determined by government legislation.

As the size and extent of generating systems increased, the problems of control and operating procedures changed significantly and a different design philosophy was required. The resulting procedures for dealing with large systems are now well established.

Most electrical equipment is designed for connection to a 'standard power system' of this type which is assumed to comply with a specified level of performance, virtually constant frequency and voltage held within the limits, which may be to statutory requirements. For such equipment, standard performance specifications, both national and international, are prepared on this basis, for obvious aspects, such as harmonic distortion, which are relevant in particular instances where the idealized conditions do not exist.

During the development of such national networks, however, there still remained a number of independent generation units to meet specialized operating procedures or application requirements, and in recent years the number and size of these has increased appreciably. Unfortunately the special requirements of such systems have often been ignored: the assumption that such systems behave in the same way as large generation systems has resulted in the utilization of unsuitable and often inefficient units and the adoption of unsatisfactory control procedures and systems. The design of any independent generation system requires a wide range of engineering disciplines, together with a careful analysis of the types and behaviour of its prospective loads and the selection of suitable operating procedures.

The simplest generating system consists of a single generator driven by some form of prime mover and supplying a single, concentrated load. Such a system can be described as a small system. It is found in practice that it is not practical to utilize standard equipment for this duty and expect performance of an equivalent standard to that of a large system. In some instances this may not be significant; thus a single electrical generator used to supply a motor-driven pump can result in appreciable variation in pump speed under varying conditions but this may not be a serious operating advantage, and the financial and technical advantages of using standard equipment may outweigh any technical operating disadvantages.

An increasing number of small independent systems are now being required to have operating characteristics as good as that of a large system and in some instances superior performance is essential. In such cases it is necessary to consider all aspects of behaviour in considerable detail to identify the unique problems involved and to determine satisfactory solutions to these.

Systems falling within the category of small systems can be of a variety of types, such as:

1 One or more generators connected to a common busbar supplying a distributed domestic or industrial load, possibly requiring special generator operating performance.

2 One or more generators supplying relatively large motor units, for example, form two-machine systems in which the individual motor and generator characteristics interact and each has a significant effect on the behaviour of the other. In addition the load characteristics may have unique characteristics and requirements.
3 Multiple generators in a compact load area, such as an offshore platform where both the above conditions may be encountered and, in addition, the unique fault conditions may impose unusual duties on the equipment.

The behaviour of all generators in the above types of system differs from that of those connected to large systems whose integrated effect presents a relatively low input impedance (i.e. approaching infinite busbars) and a relatively high mechanical inertia (i.e. virtually instantaneously constant system frequency), in relation to a local generation package connected to the system.

Intermediate conditions obviously occur but they can usually be regarded, reasonably accurately, as a small system connected to a large system through some form of simplified network.

Generation by private companies is increasing rapidly for a variety of reasons, and while only a few installations of individual generators for domestic use are now required, an increasing number of sets are being installed to ensure continuity of supply in the event of mains failure on installations where some power is essential, e.g. those serving hospital operating theatres, large computer installations or sophisticated communications systems.

Combined heat and power schemes (CHP) employ private generation which may be interconnected with a mains supply or operated in island mode. Other small systems are used on board ships for domestic power and possibly propulsion, and may be used for special duties such as cargo pumping or well-drilling on vessels. A related field of application is the offshore platform, which has similar requirements and also provides scope for supplying power to satellite platforms or subsea installations, by means of subsea cables, where local generation might be uneconomic.

The privatization of the electricity supply industry in the UK is likely to result in the installation of a considerable number of generators which will involve the problems related to small systems.

On all but the simplest of small generation systems it is now customary to centralize the control function and to use a power management system (PMS) to integrate all essential controls at a single location, providing the facility for part or full automation, as required.

With systems involving CHP it is desirable to coordinate the electrical and thermal function controls. The PMS can easily be extended to include the control of mechanical plant involved in the thermal energy cycle and act as a total energy control system (ECS). It is usual to include the economic factors relating to full utilization in such a system; these can include such items as tariffs with any time variation, fuel costs, which can be adjusted as required, and the effects of using alternative auxiliary systems where these are available.

1 Generator performance

The purpose of an electrical generator is to produce electrical energy to some load but it must do so within specified operating limits of frequency and voltage. When it operates in **island mode** it is responsible for determining these parameters but when operating in **parallel mode** with other generators it only has limited control, in general, and the required mode of operation can be as significant as the electrical load being supplied.

The load behaviour affects the generator requirements but the environmental conditions, the characteristics of the prime mover and the generator control systems are all relevant to the actual generator performance, during steady duty or rated condition and also during transient, subtransient, pretransient and all dynamic conditions encountered. The actual design of any generator, therefore, has to make allowance for all these factors; however, before considering generator performance it is necessary to consider in detail the relevant aspects of the following factors:

1 Significance of prime mover.
2 Electrical loads and systems.
3 Environmental effects.
4 Generator control systems.

Significance of the prime mover

The purpose of a prime mover is to convert energy in one form into rotating mechanical power to drive the electrical generator, and the important characteristic is the torque–speed performance as a function of load absorbed by the generator. The inherent (or design) performance of an engine is affected by several secondary factors such as variation in fuel or energy medium, or environmental effects such as ambient air. The resultant performance will seldom result in a satisfactory generator performance to meet the requirements of the electrical load and a control system is added to ensure reasonable compatibility. This may consist of a simple governor or be a more complicated energy-regulating system and it differs significantly between types and sizes of prime mover.

The prime mover and generator form an integral package with mechanical connections between them and the wide range of possible features associated with this interaction requires the generator to have mechanical features which can affect its electrical performance.

Prime mover

The type of prime mover chosen for any generator must meet the following basic requirements:

1. Have a rated continuous running speed which ensures that the generator will produce its desired electrical frequency, if necessary by using a gearbox or other speed-changing device between prime mover and generator, and possess some means of adjusting the speed.
2. Use a suitable, available, economic fuel, and have an effective fuel supply system including any necessary filtering, pressurization etc., with a standby alternative and a suitable changeover procedure, if desired.
3. Have acceptable initial and operating costs, including running costs under actual operating conditions of load, fuel and environment and allowing for effects of running efficiency, heat/energy recovery, and also expense involved in supervision, erection and maintenance.
4. Be capable of operating satisfactorily and safely in its specific environment, such as available cooling media, weather, hazardous area, marine or other contaminated location, and adverse effects of noise, vibration, exhaust etc.
5. Have a competitive first cost in relation to number and size of units selected, including any auxiliaries required, together with structural or installation requirements.

In addition to these basic requirements, however, the following technical aspects must also be satisfied:

6 The complete interconnected shaft system, including all rotors, gears, couplings, etc., must be designed to meet all operating torque and speed conditions for steady-state load conditions and also any cyclic, torsional, transient and pretransient electrical short circuit duties, as well as starting and accelerating conditions. It must also be compatible with all vertical and horizontal displacements, deflections and line-ups, including axial effects, and their resultant effects on the components of the system and its supporting structure.
7 It must also be suitable for any abnormal torque conditions or irregularities due to maloperation, cyclic and torsional effects and resonances, whether mechanical, electrical or a combination.

It is also necessary to incorporate a suitable speed-governing and compatible fuel supply control system, which must meet all the following conditions:

8 Provide a suitable steady-state and transient speed change control allowing for the specific fuel system available and auxiliary supply arrangements, including, if required, a suitable on-load fuel changeover facility.
9 Provide a suitable steady load–speed characteristic, droop or isochronous as required, with acceptable levels of error, fluctuation and repeatability, with adequate range of control facilities and maintenance requirements.
10 Prevent undesirable interaction of multiple sets in parallel, whether similar or dissimilar, through any energy medium whether electrical, mechanical, fuel or any other, and avoid any interactive electromechanical resonances.

General

While the above summary is useful as an aide-memoire, in practice it is usually found that there are other factors which may have an overriding influence on the optimum choice of plant. Thus prospective purchasers may have an aversion to a particular type of machine, or a supplier, or there may be national or political reasons for supporting home manufacturers' products in preference to those from other countries.

Some units have been specifically designed for particular applications and match the requirements more closely than others; some, by reason of long proven service, have an obvious advantage over newer designs which may appear technically superior on paper, but do not have a proven service record.

It is necessary, therefore, for a system designer or project engineer to take into consideration such factors, as it is not usual for the choice of every feature of a unit design to be unrestricted.

Usually, any investigatory study for a suitable generation plant will result in a final choice between significantly different concepts: for example, a low-speed diesel-driven generator versus a high-speed gas turbine driving a generator through gears. Nearly all the items listed in the above summary have significantly different effects on such options, and the final assessment as to which unit is more suitable for a particular installation may have to involve completely different concepts which are not normally of much technical significance. The psychological reaction of operating or maintenance staff can also play a big part in such choices, and the necessity for providing special training or educational courses may have to be included in the final cost factors.

The primary function of a generator prime mover is to convert energy from fuel or other source into mechanical movement which can be used by the generator to provide electrical energy to the load. The most common source of energy is fossil fuel or water power; nuclear, wind, wave power etc. are also used, but to a lesser extent. Fuels such as oil or gas can be used directly in engines, but they can also be used in the same way as coal or wood for combustion in a boiler to raise steam which can drive other forms of prime mover.

There are distinct types of prime mover, those that have a pure rotational movement, such as a steam or gas turbine, and those such as a diesel or steam reciprocating engine; the method of producing torque to drive the generator results in very significant advantages and disadvantages of these alternative types.

Development over the years has resulted in each type being employed for specific purposes and for the range of power and speeds where it offers significant advantages. Thus, in general, diesel engines and hydraulic turbines are directly connected to electrical generators, whereas gas and steam turbines, which operate efficiently at high speeds, are usually connected to generators through speed-reducing gearboxes.

Special applications may justify more unusual arrangements to match particular requirements, but the same basic conditions require to be met if satisfactory generator performance is to be obtained: in some instances mixed types of prime mover can be incorporated with advantage, but it is always necessary to evaluate not only first cost of special arrangements but hidden costs related to expensive or difficult maintenance and possibly operator training to handle complicated plant.

Characteristics

The main difference between reciprocating and rotary engines is that the former generates a fluctuating torque while the latter is inherently a uniform torque. A variety of features have been used to minimize this difference, such as the use of multiple cylinders which are utilized in a time sequence to produce a greater number of torque pulsations but of a smaller magnitude. The addition of a flywheel can also reduce the effective output torque by acting as an energy damper and converting torque pulsation to speed pulsations of small magnitude.

In many instances this torque–speed pulsation has a negligible effect on generator performance. However, it can be very significant if the frequency of the pulsation, which naturally results in a pulsation of the electrical energy output, coincides with a critical response frequency of some electrical load, such as the phenomenon of lamp-flicker, or the more common condition where the frequency of pulsation is such as to excite an electromechanical resonance or even a purely mechanical resonance in some part of the associated structure.

The electromechanical resonance effects, although originating in the prime mover, can be propagated through the electrical system and can affect other units on the system. This is of particular importance when running generators in parallel where their prime movers are of the reciprocating type.

The factors that determine the power output of different types of prime mover are quite different: operation at low power can cause serious deterioration in a diesel engine, whereas it can often provide considerable overpower for a short time; rotary machines do not exhibit this to the same extent. Since gas turbines utilize large volumes of air for combustion, their power output capability increases considerably at lower ambient temperatures.

The quality of fuel used can also affect the capability of any prime mover, quite apart from the possible damage that low grade or contaminated fuel can do.

Different types of prime mover are designed for specific types of duty and the basis for their rating takes this into account. For example, industrial-type gas turbines are heavier and more robust than the aero-derivative types and are intended for continuous rated operation over considerable periods, whereas the lighter, more compact aero-derivatives have advantages in short-term emergency operations.

It is essential to evaluate all aspects of power capability of prime movers together with the other relevant factors prior to selecting the most suitable prime mover.

Speed control

Power output is a combination of torque and speed, and the speed characteristics can be as important as the torque capabilities. Ideally, a generator should be driven at constant speed to produce a steady system frequency. However, this is difficult to attain and a more practical approach is to keep the speed variation within fine limits. The range permitted depends on the actual load requirements and hence different forms of speed control could be required to meet all conditions. The simplest form of governor uses speed sensing; that is, when the speed changes, a control action takes place (such as changed rate of fuel supply) to try to restore the original speed. If this system is successful and the speed is restored to the same rated value for any value of load, this is described as isochronous control, but there are transient deviations while control action is taking place.

The usual behaviour of a prime mover is to lose speed when load is increased, since the energy input, e.g. fuel, remains constant and kinetic energy is used to supply the extra load. If fuel rate is not changed, the set runs at a torque and speed compatible with this energy input as determined by the engine's inherent characteristics. The rate of response of the governing and fuel supply system determines the rate at which increased energy can be supplied, initially to match the load requirements and then to restore the lost kinetic energy. When load changes are gradual, the transient speed dip will be small and disturbance minimal; when sudden changes of large amounts of load occur, the transient speed disturbances will be greater and their magnitude will be determined by the rate of response of the control system.

More sophisticated control systems can be used to give closer speed control and faster response by sensing rate of change of speed or by measuring the actual electrical load demand. However, very fast control does present problems under some conditions. A compromise can sometimes be reached by having two rates of response, one for small disturbances, speed change or load change, and a much faster one for large disturbances. Rates of response are discussed later.

In the past it was customary to use droop speed control, i.e. the steady engine speed was lower at higher values of load, and this is shown in Figure 1.1. The characteristic shown indicates a linear relationship between speed and load; although this is not usually attained in practice, the slight deviation does not affect the operational behaviour where linearity is assumed.

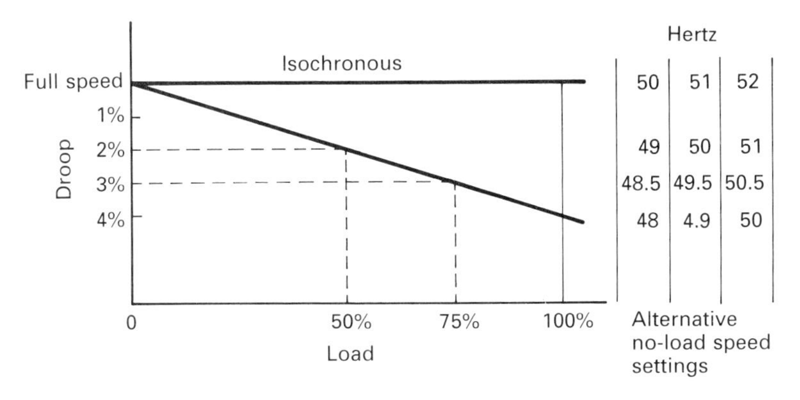

Figure 1.1 *Alternative speed–load characteristics*

The characteristic is defined by the no-load speed and the droop in speed from no load to full load. A value of 4 or 5 per cent is commonly provided for the droop speed related to the no-load speed. With a synchronous generator this characteristic can be converted directly to frequency. Figure 1.1 shows three ways in which this can be done. If the no-load value is set at 50 Hz and the droop is a nominal 4 per cent, then the frequency at full load will drop to 48 Hz. Alternatively, setting no load at 52 Hz gives a full-load value of 50 Hz. In practice, most generators do not operate at either extreme of load; a setting of 51 Hz at no load, giving 49 Hz at full load, would give 50 Hz at half load and, assuming that load variation was between 50 and 75 per cent, the frequency would only vary between 50 Hz and 49.5 Hz. This range would be reasonable for most generators operating individually.

When two generators are operating in parallel with identical characteristics and speed control systems they should theoretically share load equally, but in practice there is always some slight difference between the two sets of parameters. As the generators are synchronous with the system they must both run at equally synchronous speed, and if there is any tendency for their speeds to differ, their relative load angles will change. The generator which tends to increase speed will increase its load angle and so input more power into the electrical system, while the slower machine will reduce its angle and reduce its output. However, since the speed control systems and machine parameters remain unchanged they must resume the stable condition of equality and, in effect, the first machine supplies power to the second deficient machine to restore equilibrium. This power circulation, which is rapidly damped following any sudden disturbance, is known as synchronizing power and assumes that synchronous machines on the same system run at the same speed. Even when the generator excitation conditions differ, this phenomenon operates. However, the amount of synchronizing power available to pull the speeds together is a function of the amounts of excitation on the two generators and their impedances; if the magnitude of the transient disturbance is large, the available synchronizing power may not be adequate and the two machines can lose synchronism.

Following any disturbance to generator speed, in addition to the above synchronizing power action there will also be a tendency for the individual speed control systems to operate, if there is a net change in electrical load, and the inherent load–speed characteristics will then determine the final steady-state condition.

Consider two identical generators A and B having similar governing systems, both having a 4 per cent droop. Refer to Figure 1.2. When the no-load speed setting is selected to correspond to 50 Hz for machine A, its characteristic indicates that at 75 per cent load it would operate at 3 per cent slip. However, if machine B had its no-load setting increased to 1 per cent then it would provide 100 per cent load at 3 per cent slip, i.e. when synchronous with machine

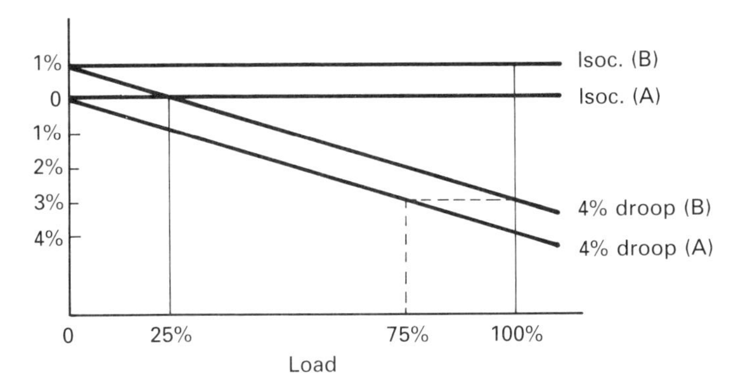

Figure 1.2 *Mismatch of parallel generators due no-load settings*

A. If the system load were reduced, machine A at 50 Hz would produce no power but machine B would be producing 25 per cent. Any further reduction in system load would result in a rise of frequency above 50 Hz and a recirculation of power between machines A and B, which is normally an unacceptable condition.

From Figure 1.2 it will also be seen that it is impossible to operate the two machines on different isochronous settings. When such control is used, one machine is nominated the master and any others are slave machines.

It is most important, therefore, to ensure that all synchronous generators have accurately set no-load speed references which are identical. However, even with this condition, as Figure 1.3 shows, if the droop differs between sets A and B there will be a load mismatch. Machine B with 3 per cent droop will take 100 per cent load, while machine A with 4 per cent droop will only produce 75

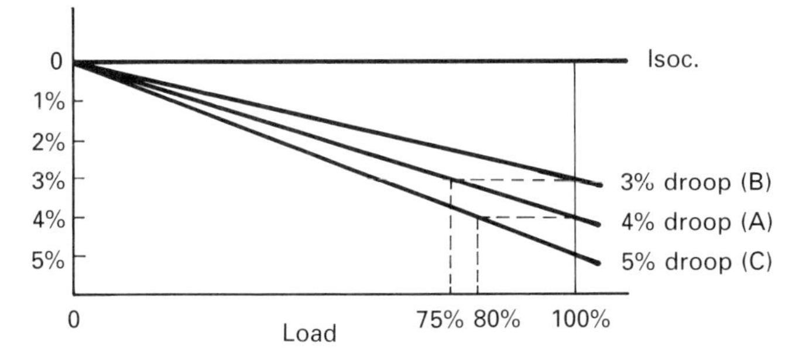

Figure 1.3 *Mismatch of generators due to different droop settings*

per cent load. Also, a machine C with 5 per cent droop will produce 80 per cent load when paralleled with machine A on full load at 4 per cent droop. Thus it is important to match both these parameters in any speed control system.

As, in practice, no two sets of characteristics are identical, it will be appreciated why droop governing is convenient: it ensures that, while machines do not share load equally, the loads are similar, and that the errors are not excessive for the accuracy of setting which can be expected and maintained in service.

However, the drop in system frequency is not always acceptable to electrical system loads and hence improved speed control systems are required.

It is interesting to consider the combination of machine A isochronous with machine B on 4 per cent droop from a 1 per cent no-load speed setting (see Figure 1.2). At 50 Hz, machine B will operate at a fixed load of 25 per cent whereas machine A can operate at any load from zero to full load. Thus we have an automatic base loading system and the magnitude of this base load can be adjusted by adjusting the no-load speed setting of machine B.

The type of speed control described thus far is essentially steady state: that is, it assumes that load changes, and consequent speed changes, are slow, and it neglects transient effects and also inertia in the control system and in the electromechanical system. All practical systems exhibit transient or dynamic characteristics which can differ significantly from the steady-state behaviour, and this is particularly important in electrical generator systems, since the electrical components have their own unique transient behaviour and respond rapidly to changes. The fuel supply system may also have a significant effect during transient phenomena, since a sudden change in demand cannot always be met instantaneously and a time-lag in response is introduced. It is also possible that interactive effects will occur on systems where several prime movers are fed by a common fuel service and a sudden increase in flow to one machine can produce a temporary reduction in the supply to other machines.

It is not possible to provide a typical transient response for all systems since the number of parameters involved is quite large and even identical prime movers can exhibit different responses if their governing, sensing or fuel systems are different.

Figure 1.4 indicates the response of a particular prime mover to three different types of transient loading with a particular govern-

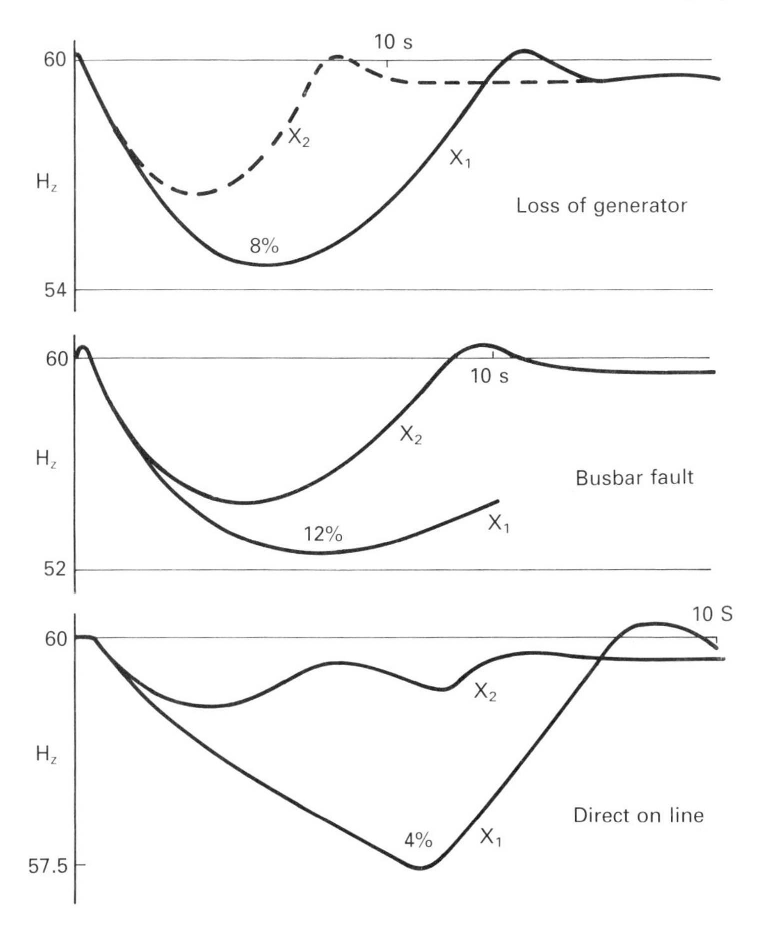

Figure 1.4 *Effect of governor response on transient load changes*

ing system X_1 and also the responses if the governor response rate is improved to X_2. The differing shapes of response are caused by the transient characteristics of the electrical load which is being changed, and the severity of the speed dip is largely determined by the relative magnitude of the load change.

With the most severe condition the speed drop is about 12 per cent and occurs about 7 s after initiation. Recovery from this is very slow and would doubtless result in instability of the system. However, using the improved governor would prevent this. Both governors would have similar steady-state characteristics.

In addition to the effects of the type and magnitude of electrical load during transient phenomena, the actual prime mover loading

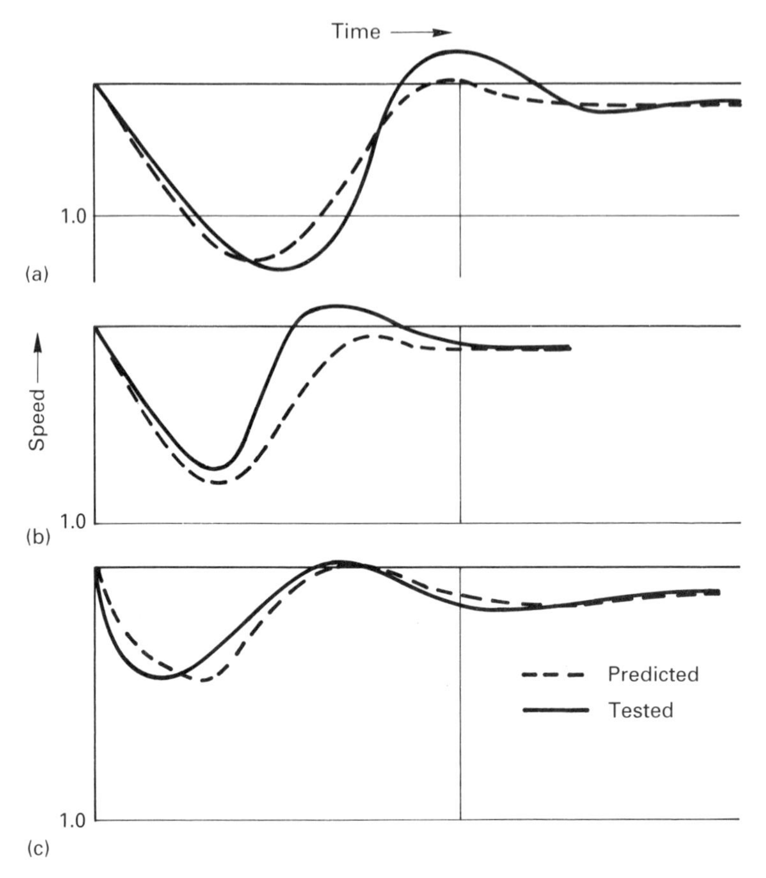

Figure 1.5 *Effect of generator base-load on transient load changes*

may also have a severe effect on the response. Figure 1.5 shows the effects of adding the same load to a gas turbine driven generator when the original loading was (a) one quarter of the added load, (b) one half of the added load and (c), (d) equal to the added load. Figure 1.5(e) shows the effect of adding the same load resulting in the machine carrying full design load, and (f) shows the same load resulting in a 10 per cent overload resultant loading.

The number of factors involved in these responses is considerable and their prediction is not easy. The solid line shows the measured response on test compared with the calculated values (broken line).

Prime movers can often have their rating increased by ancillary means such as air blowers or superchargers on diesel engines. It is

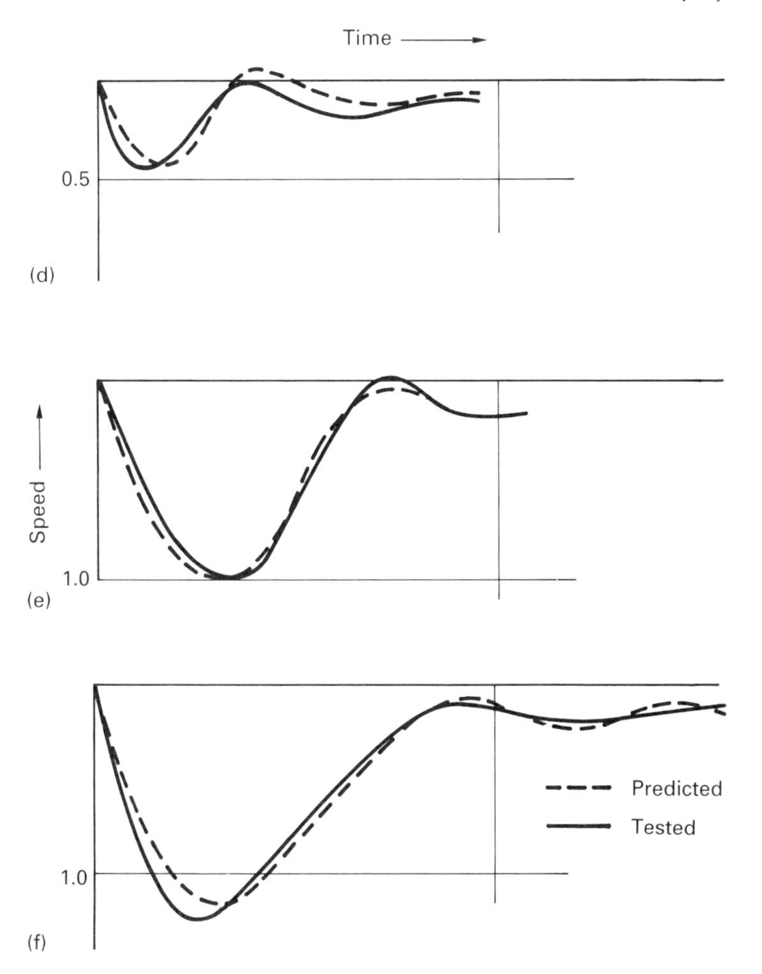

customary to control these devices to give optimum performance over the full range of possible loading. Thus at low loads the blower speed can be reduced to match engine requirements, and when transient loads are applied, time is required to accelerate the blower to the required speed and operating pressure. This produces a similar range of responses to those indicated for gas turbines and can have a significant effect on system stability during transient load changes. The transient response characteristics of any prime mover are complicated, involving a large number of parameters, and not always easy to predict. Thus, when generating sets are being operated in parallel, it is not only necessary to match the steady-state operating characteristics but the transient behaviours must also be compatible. If the dynamic characteristics are significantly different either in speed change magnitude or time, the sets will interact and circulate energy between them, and if the magnitude of this is excessive or its frequency excites a natural resonance, the sets can become unstable and may actually produce instability of the complete system.

Matching the transient behaviour of duplicate generating sets with identical governing and fuel systems can be done quite quickly if suitable loading is available to simulate necessary conditions, and a recording facility can identify the function variation during load changing conditions.

However, when it is necessary to operate dissimilar sets (with different types of prime mover, of different size or speed of operation, or with governing and fuel systems that are not matched), it may not be possible to ensure that they will remain stable during transient conditions even though they can operate satisfactorily under steady-state conditions. Adjustment of rate of response can be used to make the recovery characteristics of the sets similar, but the magnitude and time of the initial speed dip is dependent mainly on the relative kinetic energy stored in the sets and this cannot be adjusted after manufacture. It is therefore essential that this feature of generating sets is fully investigated before plant is installed in any system.

Other control modes

In an effort to improve transient performance, governing systems have been produced which do not rely on prime mover speed as the control signal. Sophisticated governors frequently use rate of change of speed or higher differentials to expedite action. Severe transients result in large speed dips and it is possible to increase the governor ramp rate, below a preset speed dip value, to give a better transient response, while retaining the more stable, slower response rate to smaller changes in load and speed.

With electrical generators it is easy to determine when the load changes and this can be done before the speed has changed and can enable the governor to pre-empt this signal and give a faster and more accurate response to the actual load change being imposed.

This type of control can be used to give isochronous speed control of a generator, and with this even transient load changes produce very slight speed disturbances. With sets in parallel it is usual to nominate one master generator to determine the precise system frequency and to control the parallel sets to share load proportionally. Such a system maintains constant speed on all drives on the system and maintains output on all induction-motor-driven devices. When used in conjunction with astatic voltage regulation on the generators it also operates all induction motors at their optimum design conditions.

The selection of a suitable prime mover does not, therefore, depend solely on its own inherent characteristics, but also on those of the governing systems and fuel systems which can be used with it, and these latter may be of prime importance in some electrical generation systems.

It is usual on a small or satellite system to do some assessment of system stability when considering modification to its generation. To carry out such a study and obtain realistic and useful information involves considerable care and knowledge and this is discussed in Appendix 3.

In addition to such transient speed control effects, which usually have fairly rapid decrements, there are other dynamic effects which can result in sustained resonances or oscillations depending on the electromechanical behaviour of the combined prime mover and generator set and the nature of the disturbing forces. The phenomenon of synchronizing power has already been described, whereby two synchronous machines can interchange power while still remaining in synchronism by means of changes in their respective load angles. Normally, the circuit resistance serves to damp out this effect quite quickly, but under certain conditions this does not happen and swinging can occur between the mechanical parts of two generator sets, sustained by the circulation of electrical power between the generators. These generators can be separated by an appreciable distance of electrical distribution network and do not need to be closely connected. Each generator set has a natural frequency of oscillation when connected to an infinite system as a reference, which depends mainly on the coupled inertia and the value of the synchronizing power, and provided that there is no exciting force at about this frequency, no adverse effects are encountered. However, if any regular pulsating effects exist in the system, such as the irregular torque produced by a reciprocating prime mover, or an electrical load such as a motor driving a reciprocating compressor, and a significant component of pulsation exists at, or near, the natural frequency of a generating set, then resonance is possible and damage can result. For this reason, speed control systems must be designed to minimize any such effects that might be expected to arise, and to avoid contributing to the oscillation effects. It may be necessary to take other action on the generating sets, as discussed below under 'Environmental requirements'.

Electrical loads and systems

Simple electrical loads

AC electrical loads are often defined simply in terms of current magnitude, in amps, but this by itself is not an adequate description, as it implies certain electrical system parameters. To be explicit the amount must be related to a complete set of system parameters defining voltage, frequency, number of phases, etc. Without further qualification it would be assumed that the system voltage and currents were sinusoidal (i.e. contained no harmonics) and that they were symmetrically distributed between the phases

Even with all these assumed supplementary parameters, the magnitude of current does not give a unique description of a load without the corresponding value of angular displacement between the phase voltage and the phase current, known as the phase angle (ϕ). The term 'power factor' is used for the value of $\cos\phi$.

Using the above parameters, a load can be defined by a current magnitude and a power factor, although it is more usually converted into a product of current and voltage, expressed as volt-amperes (VA) and power factor ($\cos\phi$). This enables the load to be resolved into two components, namely VA $\cos\phi$, known as watts or electrical power or 'real power', and VA $\sin\phi$, known as VAR or reactive power. Since A $\cos\phi$ is the component of current in phase with the voltage, this part of the load can be regarded as a resistive component R and the power W equals I^2R.

Similarly, A $\sin\phi$ is the quadrature component of current and can be regarded as a reactive component, being regarded as inductive or capacitive depending on whether the current lags or leads the voltage by the phase angle ϕ. See Figure 1.6(a).

The total effect of two loads on the same system cannot be obtained by adding the magnitudes of their currents unless they both happen to have the same power factor, and it is usually necessary to add the respective resistive and reactive components of load to derive the components of the correct total load, and hence

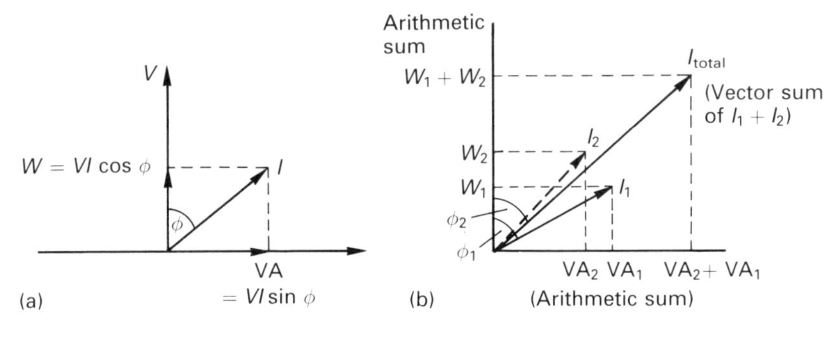

Figure 1.6 *Vector load summation*

the new current (or VA) magnitude and power factor. Figure 1.6(b) illustrates such a summation.

This same procedure can be used for any combination of loads on the same system and complying with the assumed parameters, and the total resistive component (or W) real power value then represents the amount of energy to be supplied by the electrical system to be converted into some other form of energy such as thermal or mechanical power. The reactive component (or VAR) represents electromagnetic energy which circulates round the system and is a prerequisite of most forms of electromagnetic energy conversion.

The above simple analysis is applicable to simple static loads with constant resistive and reactive values but it can also be used as a simple representation of more complicated, variable or dynamic loads provided that the necessary correct procedures are followed during the analysis.

Practical electrical loads

Any electrical load can be regarded as a static load at any one point in time and can be treated as described above, but unless all components are time-invariant the total system load will also vary with time.

Even simple, constant-value static loads have to be switched on and off; hence the total will change with time and, while the total 'connected load' can be defined, this is often appreciably different from the 'normal' duty. This deviation is identified as a 'diversity factor'.

However, in practice, even pure static resistive or inductive loads are never encountered and each load has a component of each. Also, a constant value is seldom encountered in practice, since resistances usually vary in magnitude with temperature, and factors such as saturation result in variation of inductance magnitudes. Thus, while simple static loads may appear simple they cannot be assumed to be so under all system conditions.

Dynamic electrical loads

When the electrical power is being converted into mechanical power, which forms a very large part of system loading, other significant factors are involved since the electrical system is now connected to a mechanical system and the behaviour of the latter system reacts on the electrical system. Thus, under steady-state conditions, a variation of the demand by the device driven by the electric motor will result in a variation of electrical power demand from the system, but since the energy conversion process requires reactive electrical power this latter value will also change. The relationship between W power demand and VA power demand in these circumstances is not a simple linear one and can vary significantly for different types of motor. Figure 1.7 illustrates a possible characteristic for an induction motor.

It is possible, of course, to establish values of W and VAR for any motor for its steady operating value and for practical purposes this can be used to integrate a total system load in the manner suggested above. This would represent a value of steady connected load assuming all loads were operating at their nominal rating and all

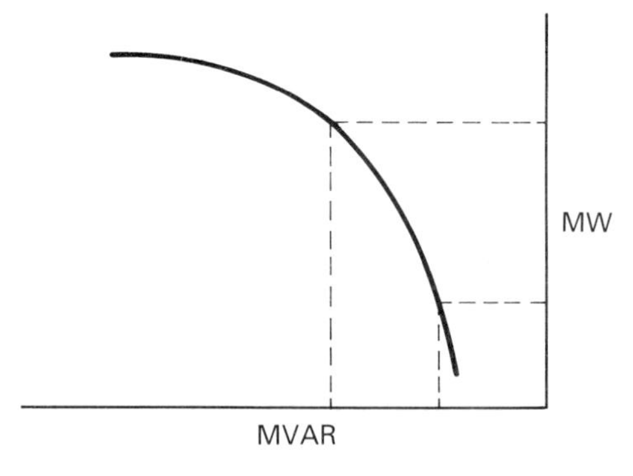

Figure 1.7 *Induction motor MW–MVAR relationship*

system parameters were ideal and constant. In practice this is not a condition encountered and it is usual to modify this value by a diversity factor to obtain a realistic total system load. Such a factor can only be determined by practice; while typical values give reasonable values for particular types of system, when any known conditions exist which are significant, it is necessary to use more elaborate and accurate methods of system load evaluation.

Effects of dynamic loads

An electrical generator supplying a motor connected to a mechanical load under steady conditions will provide reactive VA to magnetize the motor, maintain conditions for it to convert electrical power input to a mechanical shaft power output which provides the torque demanded by the mechanical load, and maintain it at the speed determined by the characteristic of the motor. Figure 1.8 shows a typical induction motor characteristic relating speed to input power.

However, any change in load torque requirements will result in an instantaneous difference between torque absorbed and torque provided and this will result in a change of speed, the amount depending upon the inertia of the total mechanical system. The change in speed will result in the electric motor changing its operating condition and will change its demand on the electrical system. This transient change in conditions will finally result in a new set of steady parameters. The change in motor demand will have another effect on the system (generator) voltage, as it will result in a change in impedance voltage drop (voltage regulation) and this change will also, in turn, modify the motor characteristics to some extent until the voltage control or regulator system restores steady-state conditions.

In many instances such transient changes are only of minor significance but there are conditions under which the system can become unstable and it is essential that they are not neglected when evaluating the load on any system.

One such condition which frequently presents problems is the process of starting an induction motor. Figures 1.8 and 1.9 indicate how the load on the system varies during this operation. The magnitude of the starting current, which is commonly six times the full-load value of the motor, can have serious effects on the supply system and must be considered when assessing system loading.

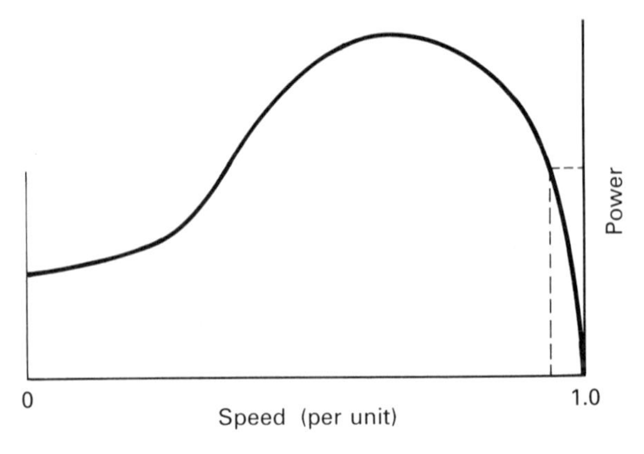

Figure 1.8 *Induction motor power–speed relationship*

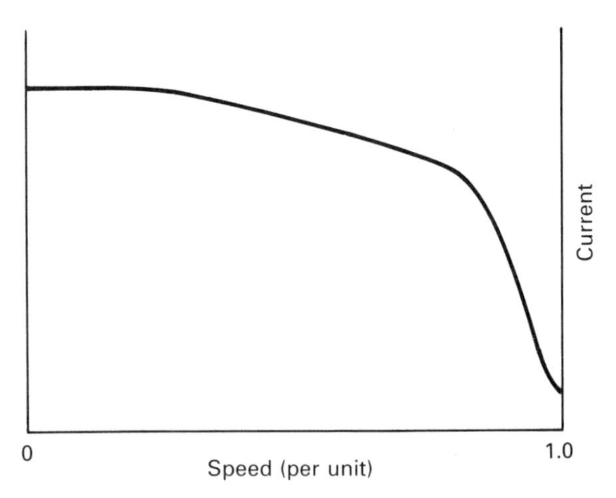

Figure 1.9 *Induction motor current–speed relationship*

Other forms of load that convert the form of energy can have equally complicated effects on the electrical system and even devices that convert electrical energy in one form to electrical energy in another form, such as frequency converters, can present a serious problem when evaluating a system load.

Synchronous motors

A dynamic load which has unique characteristics is the synchronous motor. Synchronous motors are similar in construction to synchronous generators: both types of machine can operate as either a motor or a generator, depending upon whether mechanical power is taken from, or supplied to, the electrical machine. The difference between the characteristics possessed by synchronous and induction generators described in Appendix 1 also apply to the corresponding motors.

Synchronous motors operate at synchronous speed, giving a constant-speed drive to the load, and they can generate lagging MVAR using excitation control in the same way as a synchronous generator. They can operate at higher efficiencies than the corresponding induction motors and have some distinct advantages for higher ratings. Their peak torque capability is determined by the magnitude of excitation applied and can be controlled to deal with short-time peaks, if required, in addition to providing lagging MVAR to the system. This MVAR can be used to provide that required by locally connected induction motors and so reduce the lagging MVAR required to be supplied by the system generators.

On large systems synchronous motors are often operated at constant excitation, which gives a fixed peak torque capability. The MVAR provided will vary with MW output but, if required, excitation control can be used to provide operation at constant power factor or to provide a constant value of lagging MVAR or a combination of these characteristics, including the peak torque required.

However, on small systems the excitation control of synchronous motors can be used to benefit the system by providing MVAR to assist the generators. When the synchronous motor is connected to the same switchboard as induction motors, this improves the power factor of the feeder circuit, which decreases distribution losses and improves voltage regulation at the switchboard. Where severe voltage regulation fluctuations are encountered it is possible to use the synchronous motor to compensate for all, or part, of these by generating lagging MVAR, within its rated capacity, using a suitable excitation system.

Another significant feature of a synchronous motor is that, in the event of a running motor being disconnected from a source of generated power, it will continue to run, operating as a generator, as long as its excitation is maintained. Together with its coupled load, it acts as a source of kinetic energy. In some types of drive the driven load may actually be capable of supplying a reverse power flow, e.g. a pump working against a fluid pressure feed, and this will result in sustained electrical generation. When the synchronous motor remains connected to a network with connected induction motors its generated electrical power will serve to continue driving these until the total connected kinetic energy is dissipated. Thus an induction motor with a high value of stored energy can make this available to the system as long as the synchronous motor can provide lagging MVAR to magnetize it by maintaining a reasonable value of voltage in relationship to the generated frequency. This phenomenon can be of importance during transient conditions.

Practical electrical systems

The above analysis is confined to a consideration of loads which meet several ideal conditions, including phase balance and zero harmonic content, and also assumes that the basic parameters of frequency and voltage are retained as constant, except during transient conditions. In practice, none of these conditions is fully complied with and the significance of these factors must be evaluated before it is possible to consider defining any electrical system load.

The electrical power output of any generator is derived from some form of prime mover which, in turn, uses energy from basic fuel (such as oil or gas, directly, in reciprocating engines or gas turbines, or nuclear or fossil fuels, indirectly, by the production of steam which is then converted into mechanical power by steam reciprocating engines or turbines). Hydraulic turbines utilizing 'water head' are also convenient devices for driving generators. All such prime movers require some form of speed control; the characteristics of prime movers, together with their fuel/energy supply and control system, impact on the electrical generation system and are particularly relevant when considering dynamic and transient conditions.

Electrical generators themselves are not ideal devices and produce system conditions which may have undesirable consequences in particular situations. When combined with the wide range of characteristics which can result from practical electrical loads, these problems can be very significant.

It is necessary, therefore, to consider these subjects in individual detail before attempting to analyse the problems associated with any private generation system.

Standard generators

It is not practicable to design every generator to match the requirement of each specific system loading. Standard levels of performance have been established which match the requirements of a normal diversification of loads for particular applications. A measure of flexibility is given to such standards by recognizing alternative levels of performance in some functions, e.g. in harmonic content or transient voltage response. This permits a better than average performance to be obtained where this particular function is of greater significance for a particular installation or system, such as one with predominant electric traction loads. These alternative standards do not, of course, cover all possible demands of every type of load. There are many aspects of electrical loads which can affect generator performance and which also interact with other loads having a point of common coupling with them. For example, every load has to be connected to, and disconnected from, the system; such switching operations involve significant transient effects during the change from one condition to another. These can include very high values of voltage or current for very short times, and these will normally be unbalanced between the phases. Other loads may produce considerable steady-state waveform distortion of currents or voltages and may also produce significant asymmetry between phases.

Large electrical systems

When the number of loads connected to form a single system is large, the probability is that these difficult and specialized functions will only form a small proportion of total effects on the system and their significance will be reduced. The magnitude of such an awkward load relative to the total connected load gives an indication of its probable effect on the total generation. It must be borne in mind, however, that these adverse effects could have significant effects locally on other adjacent loads.

Where systems consist of a great number of individual loads, it is probable that there will also be a large number of generators operating, and it is on such systems that operation at constant generator frequency and voltage can be expected and the use of standard generators can be contemplated.

Two-machine systems

When the number of loads on a system is reduced to a level at which the magnitude of the special function requirements becomes significant in relation to the size of generator, the idealized or standard conditions can no longer be assumed. A different design concept must be used for such small systems, or 'two-machine systems'.

In a number of installations a single generator is used to supply all the electrical load, which may consist of simple static loads or of one or more dynamic loads. The extreme case, known as a two-machine system, consists of one generator supplying one electric motor, as for example on an electric ship propulsion drive. Consideration of this case covers others involving more complicated arrangements of generator and motor used for a similar purpose.

In such systems the behaviour of the drive, whether steady state or dynamic, is directly related to the design of a suitable generator. However, the performance of the motor interacts directly with the generator, since they are of a comparable size, and the motor behaviour is also dependent on the connected ship propeller, whose behaviour in turn involves the action of the ship hull dynamics and the hydraulic effects of the sea on the propeller. Thus the generator load is a system whose behaviour is dependent on vessel design, vessel operating procedures and sea and weather conditions, and these may vary appreciably between steady and transient conditions.

At the other end of the system chain, however, the generator is driven by a prime mover, possibly a steam turbine supplied by an oil-fired boiler, and the behaviour of this component of the total thermodynamic–electromechanical–hydraulic system must be treated as a component of a single system of which the simple electric generator–motor link, that is the electrical system, is only a small part.

Such an electrical system will obviously not behave in the same way as a large system and the basic functions of constant frequency and voltage cannot be assumed. Likewise, all the special factors referred to previously will be of significance to the system, since the motor size is comparable to the generator size.

All two-machine systems require to be designed as a component of the total installed system design if satisfactory operation is to be expected from the total installation.

Small systems

In a greater number of installations a group of generators feeds a local electrical system comprising a variety of loads, as on offshore oil production platforms. Such systems also require a full study of all characteristics of the integrated loads and all their interactive effects, since the compactness of the total system, and the fact that some unit loads may approach the unit generator size, mean that many of the facts ignored in large systems can be of prime importance to safe and satisfactory operation.

With such systems it may be possible to utilize standard components, but this is not often found to be desirable. Until all the design factors have been correctly assessed, it can be dangerous to make simplifying assumptions. Such systems require the same approach as a two-machine system.

Satellite systems

Another common type of system is the satellite system, in which a small system is connected to a large system through a significant impedance such as a transformer, cable, or transmission line which tends to isolate the load characteristics of the local generation plant. Such a system may be confined to a single process plant which has installed its own generation plant to provide a high degree of power availability for an important continuous process. These systems often form part of a CHP system. It is also quite common for such systems to be capable of 'island operation', i.e. operation independent of main supply, at least to the extent of providing the essential part of the process line with sufficient power to maintain production. In island production such systems revert to the 'small system' category and require to be designed and operated as such. When systems are operated in satellite mode, connected to the main supply system, the control procedures require to be altered. Such operation must be investigated to ensure that it will meet all requirements, not only of the plant itself, but also of the supply authority that is supplying part of the power requirements of the site.

Independent generation

Independent generation can be defined as electricity generation by a unit which is not owned by the same organization that is responsible for large, widespread power distribution systems consisting of multiple generation stations and distribution networks.

The term 'private generation' has a specific connotation in the UK following the reorganization of the national supply system and the passing of the Electricity Act 1989, which permitted organizations, other than those concerned with the operation of existing plant, to build and operate generation plant connected into these systems. Such independent generation must of course be designed and operated in a manner compatible with the existing system and must meet the same criteria as a similar generation plant owned by the owners of the main system.

However, independent generation also includes a range of small two-machine and satellite systems. It is essential that careful consideration is given to the design of such an independent system, particularly to the choice of a suitable generator, since the technical problems may be vastly different from those encountered in a large generation unit closely coupled to a large generation/distribution system.

An independent generator must have characteristics suitable to all its possible operating functions, and must be connected into the system and controlled in such a manner as to provide the overall requirements of the connected loads. Such requirements may include unique safety and protection systems and, often, interaction with local or process control systems. Any interconnection with other generation systems must be analysed carefully and suitable parameters chosen for both the generator and any associated control systems. Most factors involved in such a study are interactive, and optimization of parameters and controls can only be determined after a detailed application study with, possibly, a compromise being made on some of the features to be provided. These factors will be considered in more detail.

Practical generator loads

The above analysis is based on idealized conditions, such as constant operating voltage and frequency and balanced and sinusoidal loading. In practice, these conditions are not met and consideration must be given to the effects of these factors on the generator. When the load comprises a fixed impedance, usually described as a 'static' load, the effects of variation of voltage a: frequency can be accurately predicted, but while the variation with voltage is linear, the effects of frequency are only significant to the reactive component of the load, and the voltage–current relationship is nonlinear. However, the majority of loads are 'dynamic': their behaviour is much more complicated and may involve integral control system functions. Induction motors are common examples of dynamic loads, and variation of voltage will result in change of operating power factor and efficiency. Simple change of frequency will result in speed change and the resulting kilowatt load demand will depend on the mechanical characteristics of the driven load. For example, a pump will produce a higher/lower pressure head, and consequent flow, depending on the hydraulic system 'impedance'. In addition, the power factor and efficiency will change. Simultaneous variation of voltage and frequency will result in a cumulation of the above factors. A state of operation that has some advantages in particular situations is one where the two values are varied in direct proportion to give constant volts/hertz. This condition enables the motor to operate at about constant magnetic flux so that its rated torque is produced with little variation of power factor and efficiency. It still results in a proportional change in driven speed, however.

Another category of dynamic load comprises variable-speed motors supplied from converters fed from the generator. The consequences of variation of generator voltage and frequency will depend on the control logic provided with the converter, which will normally operate to maintain the motor at a stable condition. In general, there will be significant power factor variation at the generator and consequent current loading.

A factor associated with practical loads is the presence of harmonics to a lesser or greater extent. Converters demand a non-sinusoidal current from the supply, and the magnitudes and orders of the harmonics vary greatly with the type of converter, its control logic and the characteristics required by its driven unit. Other loads have quite small distortions which by themselves would not present very serious problems. However, it is the total of the harmonics, which can vary significantly with kilowatt power demand, that finally determines the effect on the generator and, incidentally, on all other plant connected to the same distribution system. This latter aspect must always be considered carefully, since it may be more important than the effects on the generators.

When harmonic currents circulate in a generator winding they produce losses and voltages which generally detract from the machine behaviour, but in addition they result in induced currents in the generator rotor which will cause additional heating and will produce shaft torque. If no 'damper winding circuits' are provided in the generator, these currents will flow in the most convenient path and the heat they produce can result in thermal damage to the motor winding insulation. This effect can be eliminated by provision of a suitable winding, but the additional losses and harmonic torques will still be produced. The magnitudes of these can then be predicted by design calculations. Asymmetrical or unbalanced loads can produce a similar effect in a generator. Unbalanced three-phase loads can be mathematically represented by three loads, each of which is symmetrical (balanced), but which have different rotational speeds. The positive-phase sequence component will rotate at the same speed and direction as the actual generator load currents in the stator winding; the negative-phase sequence component will rotate at the same speed, but in the opposite direction; the zero-phase sequence components will remain stationary in space. Thus these last two components will affect the rotor of the generator in the same way as harmonics, producing losses and torques.

All these load characteristics exist to some extent and must be superimposed on the various types of loads discussed previously to determine the actual duties of any particular generator.

Environmental requirements

In addition to supplying the electrical loads connected to its terminals, a generator has to meet environmental requirements which relate to the effects of the environment on the machine and its prime mover, and also the effects of the generating set on its environment. These latter are becoming more significant as a result of increasing public awareness of possible damage to the environment in its widest sense.

Every electrical generator requires a prime mover and the interactions between these two units are considerable and must be considered in detail. However, the combined unit, as a single generating set, has other interactions with the environment through its mechanical structure, the electrical system, the surrounding air or other media, or the ancillary services such as fuel, cooling fluid etc. These interactions are all bidirectional and the effects may be equally significant in either direction.

Prime movers produce vibration, noise and heat to a greater or lesser extent, and can also cause pollution due to leakage, spillage or exhaust gases. Precautions may have to be taken to ensure that these effects do not reach unacceptable levels, either as may affect the generator itself, or in their impact on the local environment and on personnel within range.

Vibration effects, including torsional effects, can be transmitted through the shaft and coupling to the generator and it is for this reason that the total shaft system should always be analysed as a single entity, including all associated bearings and lubricating systems. These effects can result in damage to the structural components of a generator: the prime mover and generator design parameters must be compatible if a useful, trouble-free life is to be expected. In addition, however, these effects can produce abnormalities in the electrical phenomena in the generator, such as unbalanced magnetic pull which can be axial as well as radial, or power or current pulsation in the generated output.

Vibration effects can also be structure-borne and can cause resonance in component items on the set itself, e.g. fan blades on a rotor, or on other plant in the near vicinity, if the set is not on resilient mounts suitable for detuning the vibration. These resonance effects can quickly result in fatigue failure and must be kept to a minimum. They are very prevalent in steel-mounted sets, such as on an offshore platform or a ship, and in such instances the use of tuned resilient mounts may be necessary to isolate undesirable predictable vibrations associated with the particular type of prime mover in use.

Noise, which is virtually air-borne vibration, can be objectionable to operating personnel and other people in the neighbourhood, and must be kept to acceptable levels, not only for personal comfort but also to avoid damage to hearing. The types of noise produced by different types of prime mover and electrical generator are quite distinct and some can produce particularly undesirable frequencies. Noise can be reduced by the use of appropriate silencers but these can be expensive, may impair the performance of the set and usually add considerable weight to the installation and increase its physical space requirements. The exhaust from a prime mover usually produces a high level of noise and, even when silencers are used, it is important that the direction and location be selected to minimize environmental disturbance. A large amount of heat is produced by prime movers and this greatly reduces their overall efficiency. Most of it is contained in the exhaust and it is quite possible to recover some of this heat, and thus raise the overall efficiency, by installing a heat exchanger in the exhaust system or by using the heat to raise steam in a waste heat boiler. The mass flow of exhaust and its temperature determine what heat recovery can be made economically; whereas a gas turbine can be used to recover a large amount of heat, a diesel engine is less useful in this respect.

In addition to exhaust heat, a considerable amount of heat is radiated from the surface of a prime mover and it may be necessary to lag the machine to reduce the heat transfer to the engine room and reduce the amount of air conditioning required to maintain satisfactory conditions for the operators. It is usually necessary to remove heat from the lubricating system and this is usually done using a water-cooled heat exchanger. Air-cooled radiators may be preferred for installations where adequate suitable water is not readily available.

Electrical generators operate at relatively high efficiency and the amount of heat produced by them is consequently much less than by the prime mover. However, the total radiated heat must still be considered when designing the engine room ventilating system to ensure that a safe ambient temperature can always be maintained. This is particularly important on board ship where the outside ambient air can attain high values, particularly in or near the tropics.

The electrical output of an electrical machine varies significantly with the temperature of the cooling medium and it is usually convenient to provide a specific means of removing the electrically generated heat from the engine room rather than relying on the air conditioning system.

Generators produce heat in their electrical windings and also in their magnetic circuits. It is usual to remove this heat by means of shaft-mounted fans blowing ambient air through the machine and back into the machinery room. When this would result in an excessive rise in temperature of the ambient room temperature, an alternative cooling system is required. The rating of a generator is determined by the safe operating temperature of the insulation in the windings, and if this value is exceeded the useful life of the winding can be reduced significantly. Such an overtemperature can occur as a result of increased electrical loading in the winding, an increase in the cooling air temperature or a reduction in its mass flow.

When it is impractical to remove all the generated heat from the engine room, a pipe- or duct-ventilated machine can be provided which draws an adequate supply of cold air from some suitable source and exhausts the heated air at some convenient location. Pressure drops in the air flow circuit may necessitate the use of a separate air circulation fan, particularly if the air intake has to be filtered or the exhaust has to discharge through a silencer.

Large generators tend to utilize a closed air circuit in which the internal cooling medium is kept separate from the external room air, and the internally generated heat is removed by an air-to-air or an air-to-water heat exchanger. Thus the only heat passed to the engine room is the much smaller amount of radiated heat and the load on the room air conditioning is significantly reduced.

Very large machines can utilize direct removal of heat by passing a cooling fluid inside the conductors, but this arrangement involves considerable cost and complication.

Enclosed machines have the additional advantage of a greatly reduced noise level, as a considerable amount of noise is generated in the air-gap and can be transmitted directly to the surrounding room if the air is freely discharged into it.

Rotating electrical machines present a double hazard to personnel by contact with rotating parts or by contact with the electrical windings. Standard forms of machine enclosure have been developed to ensure safe operation consistent with satisfactory cooling, and an IEC classification has gone further and identified categories of enclosure which protect the machine itself from ingress of dangerous objects and materials. This simplifies the choice of a suitable machine enclosure to meet specific environmental conditions. This classification supplements one which defines corresponding cooling systems for the generator. These systems, however, only deal with the most usual environmental conditions and it is necessary to supplement them when dealing with special conditions such as operation in hazardous atmospheres or contaminated regions or for machines installed outdoors in severe weather conditions.

The effects of contamination can be reduced to some extent by special surface treatments of the machine structure and windings but this is only applicable to a limited extent. It may be necessary to enclose the electrical machine totally or to use a pipe- or duct-ventilated enclosure. Such a machine gives good protection of personnel and also of the machine against contamination by solids such as dust, sand, or liquids that are not directed under pressure. Total enclosure, of course, prevents the normal machine cooling process using the ambient air and a heat transfer system is required. Air can be circulated normally within the machine enclosure using shaft-mounted fans, and the generated heat can be transferred through the surface of the machine enclosure to the surrounding ambient air or other medium. This form of transfer is relatively inefficient, and for larger machines it is usual to incorporate an integral heat exchanger within the enclosure, either air to water or air to air, where the secondary coolant air may be driven through the exchanger using an external shaft-driven fan.

These totally enclosed enclosures, however, have limitations as they cannot be hermetically sealed and the pressure differences between the inside and outside of the enclosure produced when running, and the reduction in internal pressure when the machine cools down, result in contaminating material being drawn into the machine. Humid air can result in condensation on the windings with consequent reduction in insulation resistance, and conducting dusts can have a similar effect. Solvents or corrosive materials can damage the insulation or mechanical structure.

One special form of totally enclosed machine is the flameproof machine, which is designed such that, in the event of an explosive gas entering the enclosure and being ignited, the result of this explosion will not propagate flame through any enclosure gaps to the surrounding atmosphere.

This simple form of enclosure is not suitable for all conditions and it may be necessary to use a pressurized enclosure such that, in operation, there is a positive pressure between all parts of the inside of the enclosure and the surrounding medium. While this can result in a leakage outwards, it will prevent ingress of dangerous material. As an additional precaution it may be desirable to fill the machine enclosure with an inert gas such as nitrogen or carbon dioxide. This involves considerable complication and cost, which may be justified by increased safety in special instances, such as where nuclear radiation products may be located nearby, or ionization may be expected to occur.

Other environmental conditions arise which may not be adequately dealt with by the above alternative enclosures in an economic manner. Thus machines may be subjected to extreme weather conditions, such as high temperatures due to direct sunshine in the tropics, or extremely low temperatures such as –40°C in arctic conditions, and it may be necessary to take special precautions either to prevent the effects being applied to the machine directly or to incorporate material throughout capable of withstanding these effects. Driving snow can also block ventilation, and outdoor machines require to use a design compatible with such conditions. Earthquake and shock conditions, such as could be produced by underwater explosions, also require a careful review of the basic machine structural design, and even mobile generation sets may have to incorporate special features to deal with the effects of shock and vibration.

In tropical conditions the effects of fungal growths or attacks by termites or other vermin can be very severe, and even in moderate climates rodents can attack insulation and similar products. Special treatments are available for machines intended for high-risk locations.

The ancillary services provided for a generator can also provide hazards, since systems such as cooling water or lubricating oil penetrate the machine enclosure and can cause trouble themselves if they leak or if they conduct contaminants into the machine. Thus it is possible for cooling oil to absorb explosive gases in the treatment plant for cooling, filtering etc., where the oil mixes with that from a common mechanical drive such as the prime mover, and this gas can be released inside the generator. Precautions must therefore be taken to prevent contamination by such means.

The electrical system to which the generator is connected also acts as a bidirectional source of contamination. Thus any harmonic voltages produced by the generator can cause harmonic current to circulate in the system, and if there are any susceptible resonant devices connected they may suffer damage or have their function impaired. Likewise, if there are loads such as converter devices on the system, these can demand significant harmonic currents from the generator, and those flowing within the generator can cause additional heating and result in damage to the machine or a derating of its capability. These harmonics can cause maloperation of sensing devices such as automatic voltage regulators, instruments and protection relays. If they are present in any significant amount, special filtering must be provided for such sensitive devices. Higher order harmonics can also interfere with telecommunications, and machines do radiate radiofrequency power. In general these are not troublesome but when a generator is intended for use in a particularly sensitive situation their magnitude should be evaluated to determine compatibility with the environment, by carrying out a harmonic distribution system study using accurate values of machine harmonic impedances.

Phase imbalance in machines also produces system disturbances and even the conductor arrangement on stator windings can result in subharmonic fields circulating in the air-gap. These can cause pulsating structural deformation which can be transferred to associated plant through the foundation or supporting structures, and when conditions are likely to be susceptible to this phenomenon the winding should be designed to minimize this particular effect.

Another environmental hazard sometimes encountered is organic and consists of damage to insulation by growths such as bacteria or fungus or the more positive attacks by vermin such as rats or termites. Suitable enclosures help to minimize these, and the choice of appropriate insulation can be advantageous.

Generator control systems

Generator excitation control

A simplified explanation of AC generator theory is given in Appendix 1 and this indicates that the only real control which can be exercised on a generator directly is the adjustment of its excitation. The excitation winding on the rotor is supplied with DC current, and the necessary voltage required is determined by the winding resistance, which varies with temperature, and any supply circuit voltage drops. To maintain a constant value of excitation, therefore, it is necessary to have some means of adjusting the voltage applied. In practice it is also necessary to adjust the exciting current as the generator load changes from no load to full load at least, and in many instances over a much wider range than this for any of a variety of purposes.

In the past the DC excitation winding on the rotor was fed, via brushes running on sliprings, from an external source of DC power. This could be from a suitable fixed-voltage DC busbar through an adjustable resistor, but this involves an appreciable loss of power and hence reduction in efficiency. Alternatively, the DC supply could be from a dedicated DC generator set with shunt field control, which is less subject to loss, but the losses involved in the driving motor and the additional cost and complication of a suitable motor supply and starter have limited this system to only a few specialized applications. A cheaper and more efficient system consists of a rectifier fed from a local AC power supply, and over the years most forms of rectifier have been used for this purpose.

Dependency on a separate power source reduced the operational reliability of a generator and it has been customary to drive the exciting power generator direct from the generator shaft. This could be an AC exciter with a separately mounted rectifier, or a DC exciter feeding direct to the rotor sliprings. Such exciters require quite low values of self excitation power and this results in better efficiencies.

The rate of increase of voltage build-up is appreciably lower with a self-excited generator than with a separately excited one and the use of a pilot exciter to supply the field current of the main exciter provides a fast and convenient system. The amount of power wasted in the regulation of the pilot exciter output is very small but the benefit to the control system is significant.

In recent years the development in rectifiers has been such that it is now practicable to mount suitable rectifiers on the main generator shaft and supply the main generator field from a shaft-mounted AC exciter either with or without a pilot exciter, which could be in the form of a permanent magnet exciter (PMG) which ensured satisfactory voltage build-up. Such 'brushless' generators are now in common use on a wide range of types, sizes and speeds of generators.

In addition to providing a controllable value of current to meet the steady-state range of operating conditions, the excitation system is required to cope with all abnormal and transient conditions to which the generator may be subjected. These include excitation build-up when accelerating a machine up to synchronous speed to enable it to synchronize with the supply, if required. When the generator rotor is running asynchronously with the supply connected to the stator, slip frequency voltages and currents are induced in the excitation system, as described in Appendix 1, and these interact with the DC current supplied by the exciter. When there are rectifiers in this circuit considerable inverse voltage can appear on these devices and they must be designed to cope with such conditions. These effects are most pronounced if the generator has 'come out of synchronism' with the supply system and can present a serious hazard to the excitation system if they are allowed to persist for any length of time.

Sudden changes of generator load also produce transient changes in the excitation circuit and the most serious of these is a sudden short circuit involving the generator stator winding when considerable current peaks are induced in the rotor winding. The excitation system must be designed to cope with all such practical operating possibilities without damage.

The excitation of a synchronous generator is required to provide the electromagnetic field which generates its voltage, but it must also counteract the demagnetizing effect caused by the generator load current which depends on both the magnitude and power factor of the load current. Thus a single generator, operating at constant frequency and constant excitation, will only produce its rated voltage at one value of load for any power factor, and for any other value its terminal voltage will change. In practice, therefore, it is necessary to provide some adjustment of excitation to enable rated voltage to be generated over the required range of load currents and power factors. This can be done manually, but continuous operator control is usually impractical, and the operator's rate of response may be inadequate to meet a particular load change requirement. An automatic compensating system can be used such that a signal based on the actual load will adjust the excitation by the necessary amount, and such schemes, although not usually providing close control, can have advantages, particularly on small machines.

However, in general, a continuously operating excitation control system is provided for all large machines and the automatic voltage regulators (AVRs) can keep the terminal voltage within close limits, say 2½ per cent of rated value, although values as low as ±½ per cent can be obtained if necessary.

With important generators it is customary to provide an emergency or back-up manual application system which can enable the machine to provide power until the AVR has been repaired or replaced, although not with the same accuracy of control limits.

Depending on the configuration of the system to which the generator is connected it may not be operationally desirable to maintain its terminal voltage constant. Most modern AVRs have facilities to give alternative control functions, by minor circuit changes, such as droop voltage control, power factor control or

MVAR control, each having desirable characteristics for particular modes of operation.

In addition to accuracy, the speed of response of an AVR system is also important. Since the control is normally error-actuated, no compensation occurs until after a voltage error is detected and hence there will always be a voltage dip, or rise, when a load changes, and a finite time before the control can restore the terminal voltage within the required steady-state limits. With slow response systems the magnitude of the voltage change is determined by the load change conditions and the recovery time depends on the control system in conjunction with the excitation system. More sophisticated systems, however, can prevent the voltage change reaching the uncontrolled value, and can give very fast recovery to normal, but involve more complication and cost and may involve problems of instability. They are only adopted where the required operational characteristics require their use.

When generators operate in parallel on the same busbar and each has its own excitation system, including an AVR, changes in load will result in generators with different characteristics taking different shares of the MVAR load, although their governing systems will share the MW correctly. In the past it has been usual to use droop control to ensure that MVAR sharing is maintained (see Appendix 1), but it is often preferable to maintain the busbar voltage at a constant value and this can be done by using an astatic control system. This senses the MVAR flow in each generator and uses any discrepancy in proportional sharing to adjust the excitation of the generators as required; via the AVR, to maintain constant voltage and also shared MVAR.

These forms of control system are applicable to small systems where the generators and their loads are close-coupled onto common busbars or through very low impedance circuits. When there is an appreciable impedance in the connections between the generators and the loads, the control problems can become more complex. It is then necessary to consider what the characteristics of the loads are and also their power requirements.

With a generator feeding a considerable load through a significant line or transformer impedance, the receiving-end voltage, at which the load is supplied, will differ from the generator terminal voltage, or sending-end voltage, by the line impedance voltage drop, which is the product of the line impedance and the load current flowing. This voltage drop can be conveniently regarded as having two

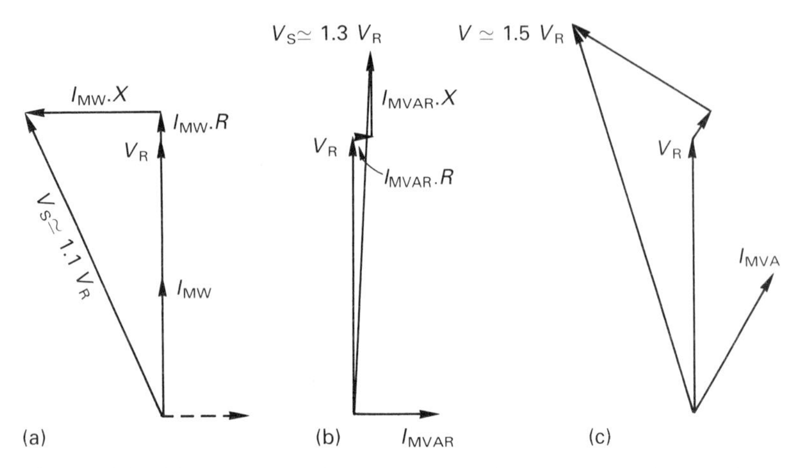

Figure 1.10 *Voltage regulation caused by MW, MVAR and MVA loads*

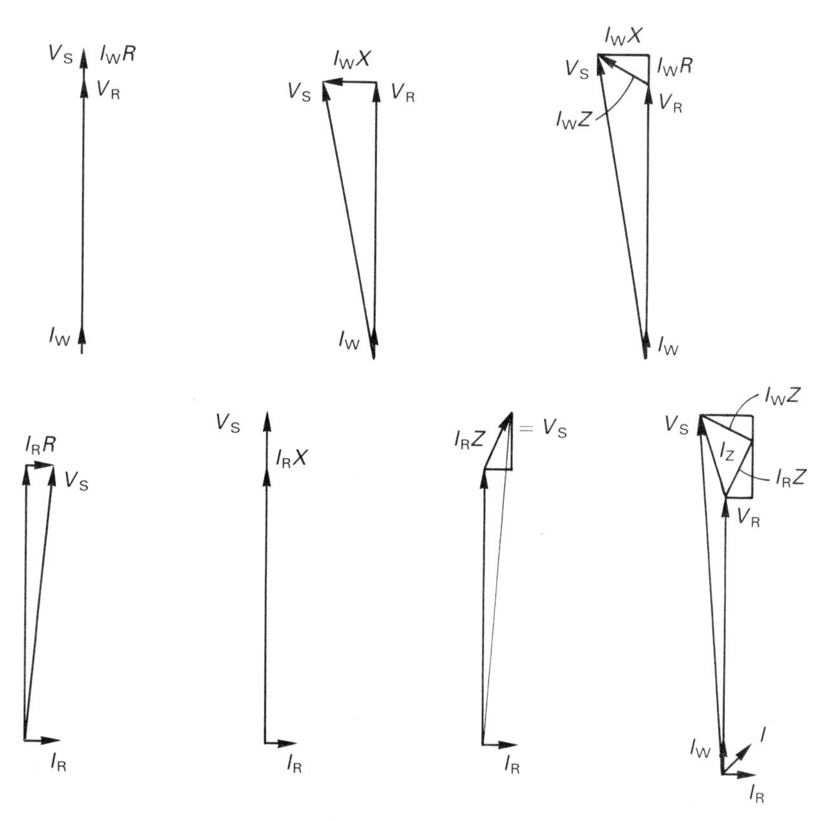

Figure 1.11 *Individual components of impedance voltage drop*

distinct components, one due to MW current flow and the other due to MVAR current flow (Figure 1.10). The line impedance is predominantly reactive and the MW drop component is in quadrature with the receiving-end voltage, producing a significant phase displacement between the sending-end and receiving-end voltages but not much magnitude difference. However, the MVAR drop acts in the converse manner and is responsible for a significant difference in magnitude between the two voltages (Figure 1.11). The magnitude of these voltage drops is variable since it depends on the actual load flow. When it is necessary to maintain the receiving-end (load) voltage within close limits, it is necessary to provide a transformer with a tap changer to give a suitable range of adjustment, or alternatively to vary the sending-end (generator) voltage. This latter option is only practicable when no load is being supplied at the generator end of the line, but it can be a useful solution in some specialized distribution systems, as described later. When the voltage at the load varies, this has a consequential effect on the actual current drawn by the load, as explained in the section on electrical loads, and this effect must also be considered when selecting a suitable generator control system.

One special application of this condition arises when a series reactor or unit transformer is connected in series with a generator to increase its effective impedance and thus reduce its fault level contribution to the system. With such an arrangement it is possible to apply simple AVR control to the generator, but by measuring the voltage on the system side of the impedance this value will be kept within the required limits. This results in the generator terminal voltage varying over a much wider range, with a corresponding range of operating flux within the machine. In addition to requiring a greater amount of excitation power, this also produces extra losses and hence higher operating temperatures when operating under full-load conditions than if the generator itself were operated at constant-rated terminal voltage.

Additional complications arise when two or more generators are connected to a system and a significant line or transformer impedance is included between them. If two such generators are controlled to constant terminal voltage then there is a restriction on the load which can be circulated between them, since the sending-end and receiving-end voltages are constrained to be equal, and the only variation possible is the phase angle between these voltages: this implies considerable freedom in MW flow but little in MVAR

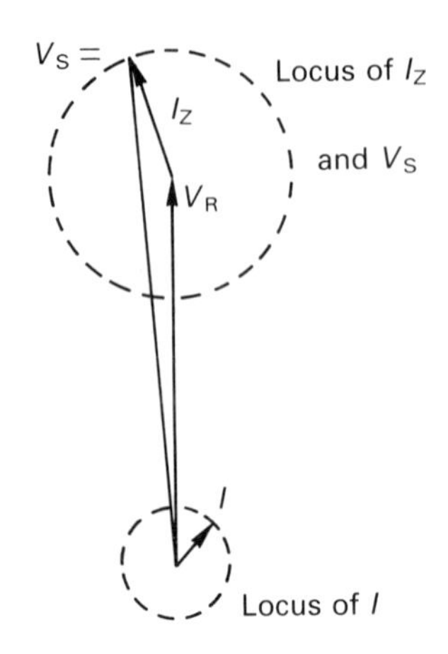

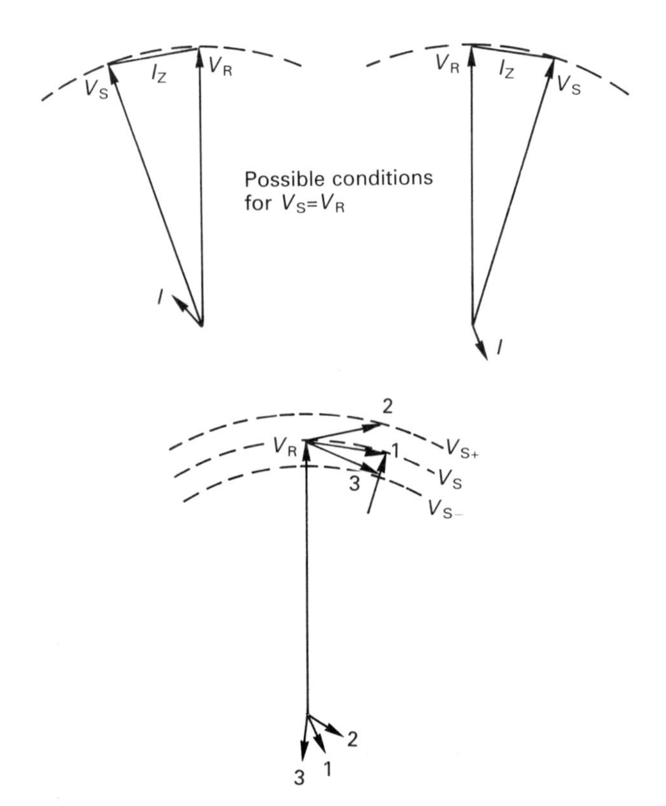

Figure 1.12 *Effect of varying* V_S *with constant I*

flow. The range of loads which can circulate is determined by this difference voltage and the impedance of the line, as discussed later (Figure 1.12).

With a simple two-generator system such as this, there are advantages in operating one (master) generator in conventional constant-voltage (AVR) control mode but using the control on the second generator to give either constant power factor (PF) or constant MVAR control, or some combination of these. In this mode the terminal voltage of the second generator will not remain constant but this will reduce the limitation on possible MW and MVAR power flows and it can be arranged such that greater MW circulation can be permitted with less MVAR circulation. Usually there is a voltage range limit imposed at ±7½ per cent or ±5 or ±2½ per cent, depending on the application. The wider the voltage range permitted, the greater the possible range of MW and MVAR circulation.

A similar condition arises when a single generator is connected to a large generation system through a line or transformer having a considerable impedance, and it is customary in such situations to modify the normal generator AVR control to PF or MVAR control. Most modern AVR units now have this optional facility incorporated, and are capable of switching back to constant-voltage AVR control in the event of the line interconnection being broken and thus allowing the generator to revert to normal operation in island mode.

These simple examples have been selected to illustrate the basic logic and to indicate which problems can arise if a suitable control system is not adopted for any particular installation. It is also important that the system selected is also capable of dealing with abnormal or emergency conditions which may arise or any special loading conditions that are likely to occur.

In most cases where such generators are installed it is usual for some load to be connected at the generator busbars, and then interconnection with other generators, and loads, results in either export or import of MW and MVAR from or to the generator. This increases the range of conditions to be considered and catered for by the control system.

Parallel operation of local generators on the same busbars can complicate the problem, but if suitable MW and MVAR proportional sharing is provided between such machines they can be regarded, operationally, as a single machine connected to the station busbar.

When more than two generators (or stations) are interconnected through impedance and when large blocks of load are also connected in this way, the determination of load flow and voltage contours requires some form of system analysis or study to obtain an accurate numerical analysis. However, some basic generator control logic must be applied using the principles described above.

The excitation system must have the capacity to enable the generator to produce its rated voltage at rated frequency and load, but in addition it must have a reasonable margin to meet normal variations of load and operation plus a reasonable margin to allow for design deviations. It must therefore be adequate to deal with the mode of control being used for the generator and this may greatly exceed the simple full-load requirement.

In addition there are special operating conditions which a generator may have to meet which impose demands on the excitation system. This may require an increased ceiling voltage of excitation which results in increased excitation power and provides a measure of field forcing, or it may require a faster rate of build-up of excitation current by reducing the time constant of the excitation loop, or more usually a combination of both these factors.

One such condition frequently encountered arises when a generator is subjected to a sudden short circuit at its stator terminal or on the nearby system, when the resultant fault current produced by the generator causes such a demagnetizing effect that the generator terminal voltage is drastically reduced and the resultant current flow becomes inadequate to operate normal protective devices. By providing such a generator with field forcing by providing a ceiling capacity of two or three times that required for the steady rated load condition, the resultant fault current sensed by the protection can be maintained at a higher value and so ensure effective relay operation.

The speed of response of the excitation system is particularly relevant when the characteristic of the electrical load supplied from the generator is particularly sensitive to voltage and it is advantageous to minimize the duration of any abnormally low voltage which can occur during load changes or when switching loads on. In some cases it is desirable to limit the duration of transient overvoltage resulting from switching loads off.

This is also very important during abnormal system conditions such as faults on the plant or distribution system which result in system voltage depression; it is desirable for the generator to recover the system voltage as quickly as possible when the protection system has isolated the fault from the residual healthy system. During such transient conditions of system voltage there are also transient changes in the system frequency and it is the combination of voltage and frequency recovery responses that determines whether the system will remain stable and normal operation of the healthy plant will be restored.

It is important to consider the generator governor transient behaviour simultaneously with the generator excitation control system, because there are certain configurations of generators and types of load which can benefit from a slower frequency recovery in conjunction with a fast voltage recovery. The optimum combination depends on the nature of the load being supplied essentially from the generator and its behaviour under such specific transient conditions.

For important installations where loss of plant can prove hazardous to personnel and expensive in loss of production, a careful analysis must be made before selecting optimum generator parameters and control systems.

Specifications

It is not a simple matter to specify a generator set and control system to provide optimum operating conditions for every installation and in many instances it is not necessary.

In many cases the generator is used solely to supplement existing generation plant and is not dedicated to a specific load. It is merely required to supplement the MW and MVAR supplied from the station busbars to an existing system which is presumably operating satisfactorily and stably.

For such an application it is usually adequate to supply a 'standard' generator, i.e. one which complies with some recognized standard specification such as BS, IEC, Lloyds etc. These incorporate the most usual limiting criteria which determine the suitability of an electrical machine for a specific purpose, based on past experience with a large number of generators used in typically diversely loaded systems. They do include optional levels of performance of some functions which may be particularly significant, but these are not related to abnormal or transient requirements and are selected to meet the user's wishes, e.g. lower than standard noise levels.

Users involved in the installation of large numbers of machines frequently require compliance with a much more strict and extensive specification of requirements which have been found desirable or necessary based on past experience in their particular circumstances. These only deal with the factors considered relevant to their particular application and may not be suitable for machines in general or used for a different purpose or in a different operational mode.

Unfortunately the existence of quite extensive standard or special specifications has led to the misconception that machines complying with them will be satisfactory for all applications and modes of operation, and this has led in some cases to inefficient, unsatisfactory and sometimes unreliable operation because of unique problems associated with particular installations.

The possible number of permutations of factors relevant to optimum choice of a generating set is very high, and an understanding of several engineering disciplines is necessary to appreciate them and their possible interaction. Without a wide experience of installations incorporating machines and plant with similar combinations of operating factors, it is not easy to ensure that all relevant factors have been adequately considered and that a suitable and satisfactory choice has been made for any particular installation.

This is particularly relevant in small installations of generators, whether in island mode or in association with other generation coupled through some impedance. It is in such cases that it becomes necessary to analyse the application requirements using the criteria described.

However, in addition to the aspects considered in this chapter there are other factors which are also relevant and which must be considered before finally determining the optimum plant to be selected for any particular installation.

2

Generator configuration

Single generator system

The simplest configuration is that of a single generator supplying a mixed electrical load at its terminals or busbars, i.e. the effects of line impedance can be neglected (Figure 2.1(a)). The prime mover must provide an acceptable speed–frequency of system to meet the load requirements, whether steady state, normal, abnormal or transient. The generator rating must meet the total specified load, but this may not be the total connected load when a diversity factor is being used. The control logic requirements will be determined by the actual requirements of the load as previously described. When details of the components of the load are not identified, a generator designed to a suitable standard specification should prove satisfactory, but the design should be adapted, if necessary, to meet any specific requirements that may be known.

In such single generator installations, problems that can prove particularly difficult are associated with loads which are phase unbalanced or which involve considerable amounts of harmonic current. However, other aspects, such as fault levels, selection of suitable switchgear and protection systems, do not usually present any problems.

When a single generator supplies a single dedicated static load, its precise range of possible duties can be specified precisely and a

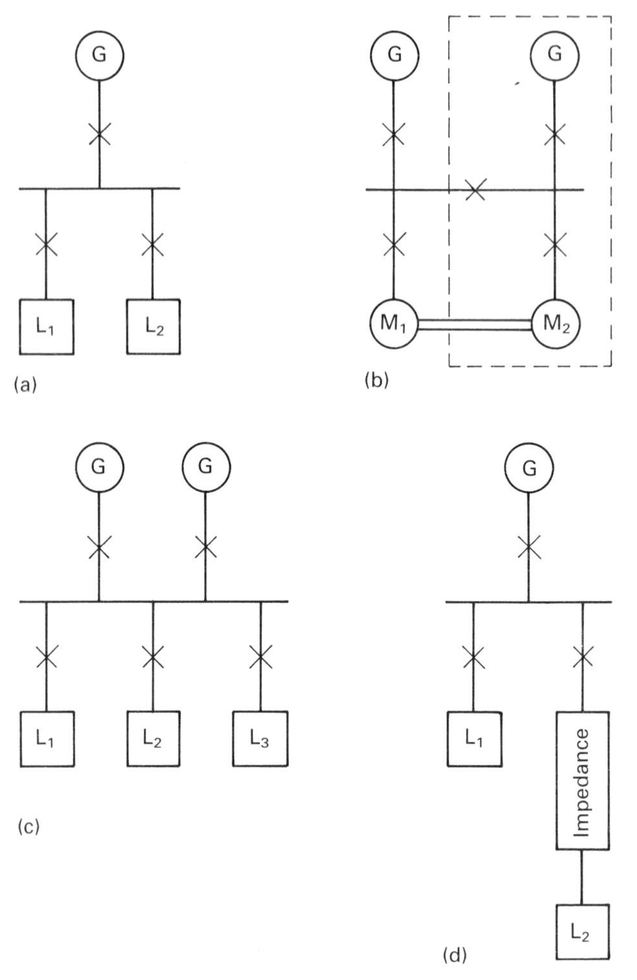

Figure 2.1 *Alternative generator configurations. (a) Single generator. (b) Two-machine systems. (c) Multiple generators. (d) Impedance-connected load*

suitable set selected or designed to meet all requirements. Thus a machine intended to supply a pulse load will have to be designed primarily for suitable transient, or possibly pretransient, parameters, and particular attention will have to be given to the dynamic electromechanical performance of the generating set, since these are the features mainly involved in such loading requirements. Fast-response control systems will probably be necessary to ensure rapid recovery of steady-state conditions following each pulse, reducing the interval before the next pulse load can be handled.

Two-machine system

A single generator supplying a single dynamic load such as an electric motor driving a specific load presents entirely different problems (Figure 2.1(b)). It is essential that the steady-state and dynamic performances of both the electrical generator and motor are matched to ensure that optimum performance is obtained at a reasonable cost. With such a two-machine system the normal control modes may be quite unsuitable and the complete prime mover–generator–motor package should be designed as a single unit to match the load–speed requirements of the driven load. The frequency control and excitation control systems can then be selected to give satisfactory performance of the load being driven by the motor.

A typical example of this is an electrical ship propulsion system, where the ultimate load is imposed by the ship propeller, which follows one of the well-known propeller characteristics, which have a common general form but where the magnitudes of the various parameters are determined by the interaction of the propeller blades with the water together with the hull contour drag characteristics. The effect of relative movement of the vessel and the sea is also relevant and manoeuvres such as turning, reversing etc. introduce another set of loading characteristics.

It is necessary, therefore, to use the prime mover speed-governing system to ensure satisfactory propeller speed control against the known load–speed characteristics, either directly if a synchronous propeller motor is being used, or in conjunction with some form of speed-adjusting device, either mechanical or electrical. The purpose of the electrical generator–motor link is to provide this control in a convenient and efficient manner.

When a synchronous two-machine system is used, i.e. with a synchronous generator supplying a synchronous motor, this link behaves in the same way as a gearbox and the propeller speed is always proportional to the prime mover speed up to the maximum power which can be transferred through the synchronous electrical link. This scheme requires a variation in generated voltage and it is convenient to control this to maintain a constant value of voltage per hertz of the generated frequency, since this results in both machines operating at constant flux and hence optimum condition of core loss together with constant torque capability. The copper losses in both machines can then be optimized by using the motor excitation control system to maintain the power factor at unity, which corresponds to minimum current required to meet the actual load demand. Such a system will, of course, be designed to meet abnormal conditions of load, within the maximum power limit of the two-machine link. The capabilities of the generator and motor are designed to be of the same order to minimize total cost for a specific value of peak load.

This is a typical case where a separately driven excitation set will be used to provide both machines in preference to two directly coupled exciters in order to reduce overall size and weight of the total installation.

A two-machine system using constant frequency generation requires some other means of controlling vessel speed and this can be done in various ways, such as:

1 Providing an electrical frequency changer or convertor between the generator and the propeller motor.
2 Providing a variable-speed motor.
3 Providing a mechanical speed-changing device between the motor and the propeller.
4 Providing a variable-pitch propeller which can vary the thrust applied to the vessel while operating at constant speed.
5 Providing combinations of these alternatives.

With all these alternatives, the generator can be regarded as supplying a single dynamic load and must be designed on this basis.

There are many possible variations of the simple two-machine system in which tandem or multiples of generators and/or motors, either duplicates or having different characteristics, can be used to extend the effective and efficient range of operating conditions and render these more flexible. Such machines should be designed on the same principles as the single machine.

It is sometimes convenient to use a change-pole propeller motor (which gives effectively a 'gear change') in conjunction with a variable-pitch propeller, which enables the latter to be run at a lower fixed speed for low powers and hence operate at a much greater efficiency. In such a case it is necessary to design the generator to be compatible with either motor winding together with the protection and control systems.

Other forms of two-machine systems exist for particular applications, but they can all be dealt with in the manner described for ship propulsion schemes.

The choice of prime mover and generator type, enclosure etc. depends on many factors, as previously discussed, and can only be correctly made when all relevant information is available. In single generator installations the most important considerations tend to be tactical rather than technical, and questions of suitable fuel supplies, servicing and spares may be overriding. In some instances reliability is the most important criterion, whereas on board ship or platforms, factors such as weight, space, noise or vibration can also be very significant. However, the satisfactory dynamic performance of the complete generation system cannot be neglected when choosing suitable plant for a particular installation.

Island generation systems

These comprise installations which consist of a multigenerator station, such as provided on an offshore platform, or in a pipeline pumping station, which may be single purpose or of a mixed application type (Figure 2.1(c)).

The design of such a system involves many interactive features, most of which can have a significant effect on each major component of the system. It is therefore necessary to use an interactive procedure if it is hoped to approach an optimum design for any specific application. Should a non-optimum system be adopted it can have any or all of the following consequences:

1 Overrun of estimated cost and increased requirement for equipment during the course of design, installation and putting to work to meet the specified requirements.
2 Unsatisfactory performance under abnormal or unusual conditions, possibly necessitating subsequent modification.
3 Unsatisfactory relationships between designer/supplier and purchasing/using/inspecting authorities, with the possibility of statutory involvement or legal action.

In many instances the supplier of electrical equipment does not have access to all the relevant data required for a composite comprehensive design and may have specific conditions imposed which prevent an optimum design being obtained. This does not relieve him of the responsibility for the satisfactory operation of the system, however, and the electrical system should meet this criterion even if the other desirable conditions cannot be met.

All the following features are relevant to the design of an optimum system and must not be ignored.

Power requirements, covering all forms of energy requirements

1 Electrical power demand in terms of magnitude and time, and both immediate and possible future requirements. Any special characteristics of supply required, such as speed variation, reliability etc. Details of levels of standby or redundancy capacity, and also dependability and control, servicing and maintenance required.
2 Energy source, including type and availability of fuel, together with supply arrangement, storage facilities, possible contamination problems and costs.
3 Prime movers, weight and cost factors in relation to possible sizes of units, acceptable mechanical features such as speed, use of gears, exhaust, noise, vibration, spares, service etc., together with all auxiliary requirements, such as possible heat recovery units, and overall efficiency.
4 Electrical distribution, including plant required, relative locations and distances between them, limitation problems such as selection of voltages.

Electrical component design, to include all equipment between the prime mover and the load

Having selected plant, after considering the above, to provide a compatible system design as a first approximation, or a number of acceptable alternative combinations, it is then necessary to consider the following features to determine a suitable set of electrical system components.

1 Generator compatibility with the selected prime mover, as regards torque, shafting and mechanical details, and with the environment. Suitability to supply the system load requirements

under specified conditions and also under conditions likely to be encountered in service.

2 Switchgear suitable for all loading conditions, normal and abnormal, and also under the most severe fault conditions. It must also meet system stability requirements and control modes, and satisfy any environmental requirements.
3 Distribution transformers, cables, lines, auxiliary boards etc. must meet all loading duties and environmental conditions and must have values of turns ratios, tapping ranges, impedances etc. to enable the system to meet all its operating criteria. The system configuration must meet any requirements for integrity, including standby/redundancy requirements, and be compatible with the selected control modes.
4 All drives must be compatible with the actual load requirements, both electrical and mechanical, and provide suitable performances during starting, speed control adjustment if any, and stability during system disturbances. They must be matched to any special operation problems which could be expected, such as resonances, vibrations etc.
5 An emergency/auxiliary system must be available for station start-up/shutdown, and also the provision of power to essential services during system disturbances. It must meet all relevant statutory regulations, and be suitable for the installation and its environment and the other factors listed above.

System control logic

This must incorporate all controls built into the above equipment as well as providing all necessary autocontrol systems required for suitable integration, and also essential supervisory and monitoring services. It must be compatible with the actual plant configuration, provide conditions which keep all items within their design limitations and make available any redundancy included in the installations. Although they are all interactive to a greater or lesser extent, the following subsystems can be designed as self-contained units, at best initially, and their interactive functions dealt with later.

Speed–frequency control

This involves the prime mover, but may also involve other devices such as electrical frequency changers or converters or mechanical speed-varying devices. In addition to controlling frequency within required steady state limits, it must give a required range of adjustment and also control transient disturbances in frequency to meet conditions of system stability. Load sharing may have to be performed by this system where machines operate in parallel, and it may be used to control redundancy in the system. Auto-starting and stopping of generation or drives may also be included and it can be used to provide economy running for some types of application.

Excitation control

This involves the complete excitation system and is used to adjust the generated voltage. It must be capable of controlling this over the complete range of working loads, and of providing satisfactory response during transient conditions to provide adequate stability margins. It may be used to maintain a constant voltage at some point in the distribution system, or provide a special voltage–load contour. Alternatively it can be used to keep the MVAR provided by the generator to a constant value or to vary it in a desired manner in relation to the load provided.

Switchgear

The generator is connected to the electrical system by means of some form of circuit breaker which must be capable of carrying the complete range of generator loads continuously but must also be rated to meet the full fault current rating which can be encountered under the most extreme conditions of system configuration and loading. It may be required to be manually operated during closing and opening but automatic opening (tripping) is required during fault conditions, and automatic closing and opening may be provided when automatic starting and stopping are required with the generating set. Several alternative forms of switch are available for this purpose, depending on the current rating and system voltage required together with the associated fault level. Manual closing may be assisted by spring loading, which can be charged electrically, or by electrical solenoids. Environmental requirements may have an influence on choice of type; for example, oil-filled switchgear can present a fire hazard and may be avoided for this reason. Some types of switch can generate serious overvoltages when being operated, and phenomena such as voltage restrikes can be encountered with some devices when the associated system parameters produce the necessary conditions.

Protection

The protection system is provided primarily to minimize damage to the generator and its prime mover subsequent to some fault condition, but it is also involved in providing system stability following some transient disturbance. It may also be used to protect plant other than electrical, and the generator protection should not be treated in isolation as the complete system protection should be regarded as part of a single integrated control system controlling the complete electrical system.

Integral control systems

As indicated in the previous sections, the various control functions all interact to some extent and it is often advantageous to integrate these in a common control mode. By combining the prime mover speed control system with the generator excitation system it is relatively simple to arrange for the automatic starting of the generating set, including any necessary auxiliaries. The set can be accelerated to a required programme, and when up to speed the generator excitation can be applied and the generated voltage matched to the supply, unless the set is being connected to dead busbars, in which case it is adjusted to the rated value. The speed control then matches the generator and busbar frequencies and adjusts the phase angle between the generated and busbar voltages. When this condition has been met, the generator stator switch is automatically closed, leaving the generator synchronized with the busbars.

This basic control logic can be extended quite simply by measuring the MW load on the generators already connected to the busbars and using the prime mover speed control to increase MW load on the incoming generator until all generators share load proportionately. When isochronous control is required, the same controls are used to adjust the system frequency to the nominated value when it diverges for any load change conditions. It is possible to share the generator MVAR loadings proportionally, either by using circuits in the AVR units, or by the provision of astatic voltage control, i.e. constant voltage operation with shared MVAR, by coupling the respective AVRs in a common astatic control loop. Alternatively, MVAR sharing can be provided by measuring the independent generator contributions and using external control logic to adjust each generator independently to give the required loading conditions.

The ability of the control logic to evaluate individual generator loading enables it to determine whether there is an adequate generating margin to meet the system load. If not, it can start another set and automatically load share it, leaving the system operating within its required working margin. Conversely, the control logic can be arranged to shut down a set if it is not justified by the current system load.

Sophisticated prediction systems can be provided to analyse load growth trends and enable sets to be put on the bars before the predicted load actually occurs.

All or some of these features are normally included in what are known as power management systems (PMS), and these can vary from very simple packages, often incorporated in the prime mover control gear, to quite elaborate microprocessor- or computer-based stand-alone systems of control.

Energy management systems (EMS)

The simple PMS systems can provide a guaranteed amount of MW and MVAR at constant voltage and frequency up to the capability of the number of generation sets available and can keep their loads proportional up to a value permitting full redundancy, if required. Thus even loss of one set will not put the system load at hazard. Alternative margins can be provided.

However, in many installations where other forms of energy are required in addition to electrical power or where an alternative source of electrical energy is available from a separate power system, it is becoming increasingly attractive to incorporate energy control within the PMS function. Where generating sets have waste heat recovery features these vary in output in relation to the set electrical loading and it is possible, by providing the correct criteria to the control logic, to let it evaluate the optimum condition of generating set loading in relation to imported electrical power, or other alternative sources of thermal energy such as boilers. Most modern boilers are supplied with microprocessor-based control units and it is relatively easy to incorporate them into an integral control system, especially if they use a common communication system. There is a multiplicity of other features which can be incorporated into an EMS and these will be considered in detail later.

Alarm and monitoring

The amount of data available to the above types of control system is considerable and it is usually economical to use the system to monitor important variables and give alarms when necessary. In some types of installation it is considered to be a hazard to use common control and alarm functions combined, but with modern systems they are usually of a duplex nature, i.e. the complete unit has a main and a standby unit, and this can restore the same level of integrity.

Such a combination enables use to be made of the appreciable computation and analytic functions now available in software and can provide much more useful information to the operator concerning the status and behaviour of all aspects of the plant.

Data handling

As a corollary to the above, with all the data and deduced information available in the control logic it is a simple matter to present what is desirable in a continuous or intermittent printout for record purposes, and it is a simple matter to include with this facility a data transmission service. Thus all necessary plant data can be presented at any required location in a form and at a time more suitable for their consideration.

This feature is very useful as a diagnostic or analytic tool as it can provide data on the sequence of events arising from any abnormality in the system and it can be used to determine the original alarm function and hence cause of the disturbance.

Emergency systems

It is sometimes desirable or necessary to take emergency action following a serious disturbance or system fault, and by combining the output from the power system protection devices it is quite practical to take the necessary corrective actions much more rapidly than would be practicable by the operator in control. The control logic can even improve on the basic electrical protection system since this is inherently designed on an equipment basis and actions follow on a cascade basis. The emergency control logic, however, can identify specific major fault conditions and initiate all the necessary actions simultaneously. This is particularly useful in complicated process plants or generation plants with much associated equipment.

Special factors

The above summary gives a good introduction to the choice of optimum plant, but in practice it is usually found that there are other factors which can have an overriding influence on the choice made, as explained in the section entitled General in Chapter 1.

Adding generators to an island system

It may be thought that a duplicate generator set can be added to an existing system without any reassessment, since all the control systems should be compatible. However, there are some factors which depend on the number of generators: it will still be necessary to check busbar loading and switchgear fault loads, as these will be increased and may exceed safe limits. Other items, such as the significance of the effective impedance of all the neutral earthing devices, must be checked and, if necessary, adjusted to give safe operation and a satisfactory level of protection. These factors also have a direct effect on the settings required on the system protection relays, since the discrimination margins will be changed.

However, when a new set is added it is not always practicable to obtain an exact duplicate in all respects, and for reasons of availability, cost, delivery time etc. it may be necessary to incorporate a completely different design of set. For economic reasons it may also be desirable to install a set using a different size, and possibly type, of prime mover and as a consequence the various new control systems will differ from the original, and the characteristics and behaviour of the generator will also be different. This may require different generator circuit breakers to be incorporated in the existing switchgear and, although it may be physically compatible, it may be different in structure, behaviour, control and protection facilities included.

During normal steady-state behaviour of the system these differences may not be very significant but it is necessary to investigate all aspects of operation because parallel operation of generators with significantly different prime movers can present serious problems. However, it is during transient conditions that differences will have the greatest effect, since the responses of the various control systems will not be identical and may, in fact, be incompatible. For example, if a small gas-turbine-driven generator of the aero-derivative type, which has a very low per-unit inertia or stored energy constant, is operated in parallel with larger industrial-type gas turbine sets, or diesel-driven generators having a relatively high per-unit inertia, their responses to a sudden system change such as loss of load or application of load will be very different. As a consequence, the small set could overload itself and trip off the bars and could result in a general shutdown.

Even when sets have electromechanical dynamic characteristics which are not vastly different, they will produce a power oscillation between themselves and this could result in power reversal which could be destructive to some types and loadings of gearboxes. It is necessary to match dynamic control characteristics to quite close limits to avoid trouble; this applies to the other control functions as well as to the speed-governing and control system, and it is usual to keep the excitation control logic as compatible as possible.

Any characteristic of the new set which differs from the original must be checked as it could introduce a new form of disturbance or maloperation into the system. This can be due to the effect of the new set on the original ones, or to some characteristic of the original sets which reacts with the new set to produce undesirable behaviour.

As an example, the addition of a reciprocating prime mover can produce a cyclic irregularity into the system which could have adverse effects on existing generation sets or loads, or the new generator could have a harmonic spectrum which contained a component capable of damaging existing equipment, or a harmonic impedance spectrum which resulted in the new machine acting as a low impedance to some harmonic already existing in the system.

Island generation with satellite loads

When a significant impedance exists between the generation busbars and some or all of the load (Figure 2.1(d)), in the form of either overhead lines, underground or undersea cables, or transformers, another control problem is introduced: the generator terminal voltage (busbar voltage) will differ from the voltage at the remote end, which supplies the load, by an amount depending on the load flowing, and it is this latter voltage that is relevant to satisfactory operation of the electrical load. Some loads can function satisfactorily with a wide variation of voltage, but others can be very voltage sensitive. Public supply systems usually have a statutory obligation to maintain the load or service voltage within specified limits which are found suitable for most normal loads, but in an independent power system this may not prove satisfactory and a load study may be required to determine a safe and acceptable range of operating voltage. In such systems it is also usual to assess the load voltage behaviour during transient load change conditions, which may actually be more important than steady-state operation for some loads.

The relative significance of the various system parameters is illustrated by Figure 2.2, which deals with the simple case of starting a motor direct onto a generator of twice its nominal rating and indicates that field forcing of at least 2.5 times rated full-load excitation would be required to maintain the voltage. However, when an interconnecting transformer is incorporated the conditions become more severe because of the impedance; while a +2½ per cent tap on the transformer enables both generator and motor to operate at rated volts on steady full load, it would be necessary to use a +30 per cent tapping to meet the same conditions when the motor is switched direct-on-line (DOL).

In practice it is usual to permit a temporary reduction in motor voltage under such conditions. By raising the motor voltage slightly above the rated value, the effective motor voltage at the instant of starting can be improved, resulting in a higher value of motor-starting torque. When the motor has run up to speed the tapping can be adjusted to give optimum motor-operating voltage. The transient behaviour of such a system is shown in Appendix 3.

When the load is fed through a single impedance the load voltage can be controlled within reasonable limits by using the generator AVR in the conventional way but using the voltage at the load, if this is convenient and practical. Alternatively, a compensated excitation system can be provided which measures the generator load, and this value in conjunction with the known impedance of the line can determine the actual voltage drop at any given time and

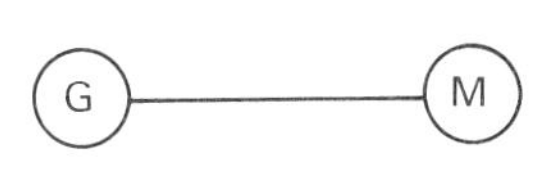

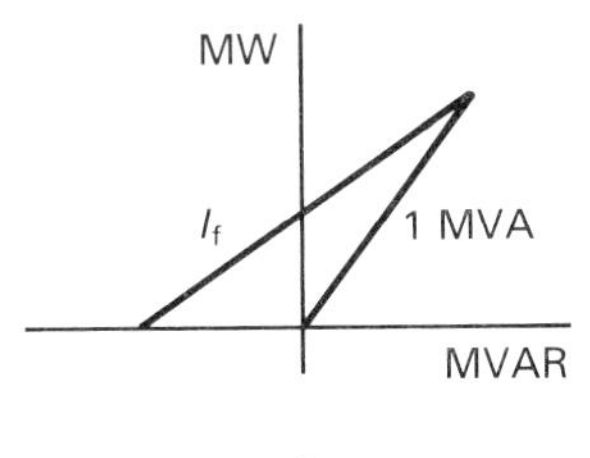

Generator rated 1.0 MVA 0.8 PF

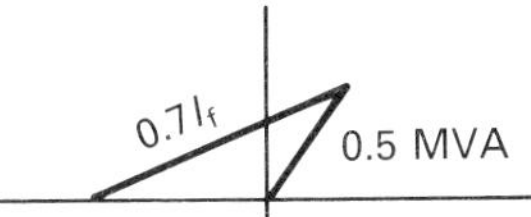

Generator with 0.5 MVA motor running

Generator with 0.5 MVA motor starting

DOL 6×FLC, PF 0.1 rated volts

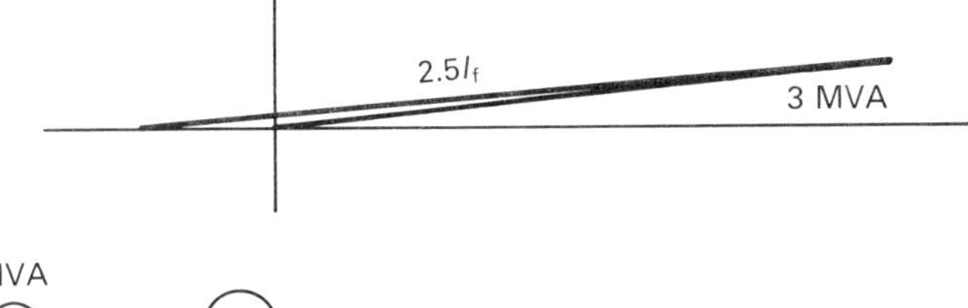

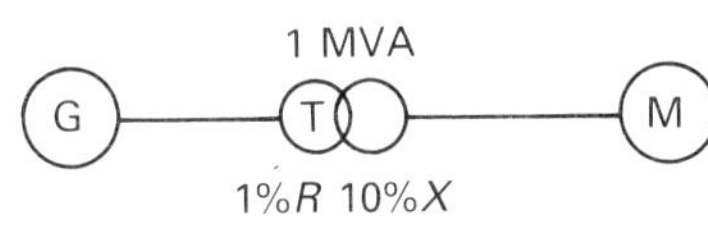

Transformer impedance volt drop

(a) With motor running: $(0.4 + j0.3)(0.01 + j0.10)$

$$= -0.026 + 0.043$$

i.e. requires +2.5% tap on transformer to maintain rated volts on motor

(b) With motor starting: $(0.3 + j3.0)(0.01 + j0.10)$

$$= -0.297 + j0.06$$

i.e. requires +30% tap

Generator loading at start is $0.233 + j2.33$
with generator and motor
at rated volts,
Transformer ratio 1:1.3

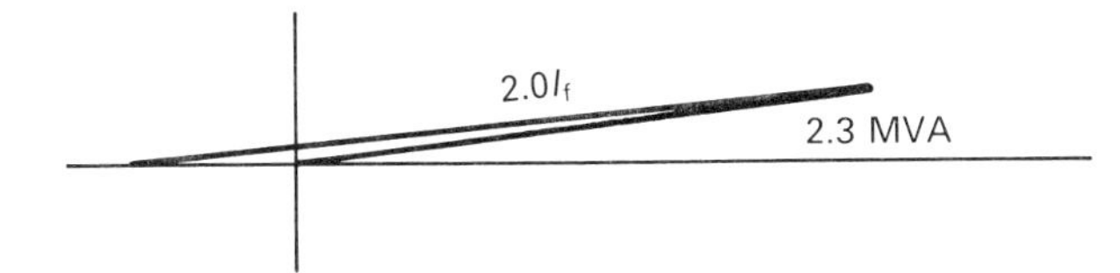

Figure 2.2 *Significance of motor starting MVAR on system*

can be used to adjust the excitation to give a generator terminal voltage which compensates for this amount. Full compensation need not be provided if the load voltage can be allowed to vary over a small range and this then reduces the range of variation of generated voltage required. These schemes have the disadvantage that the generator voltage is required to vary over a range and at higher values than the rated value, this causes increased losses and heating and may require increased excitation power.

When the line impedance and load magnitude produce a voltage drop greater than can be provided by variation of the generator voltage, it becomes necessary to use another voltage-adjusting device in the system. This commonly takes the form of a power transformer located near the load and, unless the load is virtually constant, it is usually provided with an on-load tap-changer with a range of output voltage to cope with the total impedance drop of the line plus the transformer itself, when carrying full load. This arrangement can also be used conveniently when the load is distributed in several discrete locations, by providing a suitable transformer with tap-changer within each load concentration.

However, when the line and transformer impedance becomes significant in relation to the impedance of the load, this simple system of line-drop compensation becomes unsatisfactory because of the increased losses in the line and the increased MVAR demand on the generator. It is then necessary to use a distribution system operating at a voltage appreciably higher than the generator voltage. This requires a step-up generating transformer in addition to the load-end transformer with its on-load tap-changer. The generator transformer can be a fixed ratio one, in which case the sending-end voltage of the line will vary with load proportionally to the impedance of the transformer, and in many small systems this may be acceptable. If optimum use of the plant installed is required, an on-load tap-changer should be provided on the generator transformer to regulate the line sending-end voltage within acceptable limits.

As the magnitude of the load increases or the length of line is increased, so the values of distribution voltage will have to be increased to keep the voltage regulation within the same limits. However, in practice it is usual to select a suitable value of distribution voltage from a number of standard values, because standard equipment is available for these and it is more convenient and economical to select an appropriate standard value for a particular system. As it is probable that the load will increase in the future, the voltage chosen is usually the higher one when the optimum choice lies between two standard values.

Such a distribution system can be developed to provide different load centres, each with its own receiving-end transformer, and by the use of a suitable automatic load control facility for the on-load tap-changer, a fully voltage-regulated system can be obtained, leaving the generator operating under control of its own AVR. The various load transformers can, of course, have different secondary voltages to meet the requirements of the local load most conveniently.

In some instances it is convenient to use unit generator transformers to produce the distribution voltage, i.e. each generator has its own transformer to match its rating. This arrangement reduces the fault MVA fed from each generator onto the busbars, but it also implies that the station busbars and switchgear must all be rated at the system distribution voltage rather than the generated voltage. With such generator transformers it is practicable to design the generator to operate over a small voltage range as determined by the impedance of the generator transformer and to use its AVR excitation control system to maintain constant busbar voltage.

When multiple generator systems are used with the above types of system all the forms of excitation control previously described can be employed. However although the excitation–voltage control logic is relatively simple, other problems such as protection become more complex and the design of a suitable system and its components must be selected to ensure maximum integrity of supply for the various possible modes of failure and system faults.

Interconnected island systems

When two independent island systems are combined by interconnecting their busbars solidly, the total system can be regarded as a single unit with mixed generation and mixed load, and the principles already described can be applied as relevant. It is unlikely that the two subsystems will be identical with regard to generation, control or loading, and it will be necessary to ensure compatibility between the different generator sets and their control systems, as discussed previously. Should the control systems prove to be incompatible, or even seriously mismatched, it may be necessary to make changes in design, control or operating logic to ensure that best use can be made of the two sets of generators.

However, when the interconnection between the two island subsystems introduces a finite impedance between the two sets of busbars, this has a significant effect on the reaction of one subsystem to control changes in the other and the overall control procedure can become quite complicated (Figure 2.3). The greater

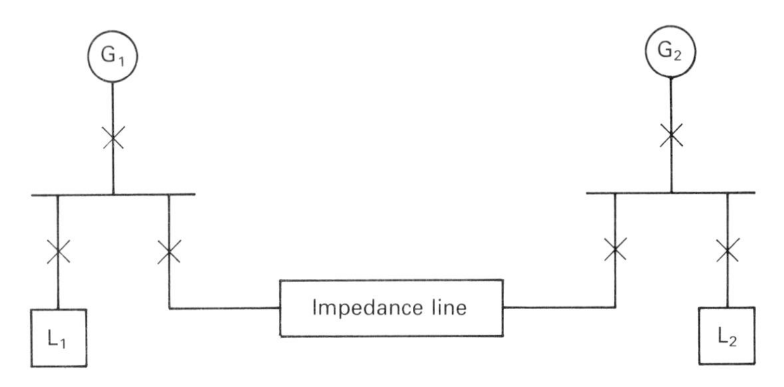

Figure 2.3 *Impedance connected generators*

the magnitude of the interconnecting impedance, the greater the complication becomes. In the event that no generation is in service in one island subsystem, the conditions revert to an island system with a satellite load, which has already been considered.

In practice, the most common reason for interconnecting island systems is to make use of surplus generation power available on one system when the other has a power deficiency, or to provide a measure of redundancy when high power reliability is required. In such instances the amount of power exchanged will not normally exceed the output of a single generating set and each subsystem will still maintain a considerable generating base which can supply the majority of its own connected load. This condition is desirable, as it improves the transient stability levels of both subsystems, but it is not always obtainable in practice.

When the two subsystems are each maintaining constant busbar voltage, there is a constraint on the current which can be circulated between them as determined by the impedance of the interconnecting link (Figure 2.4). As the amount increases, its power

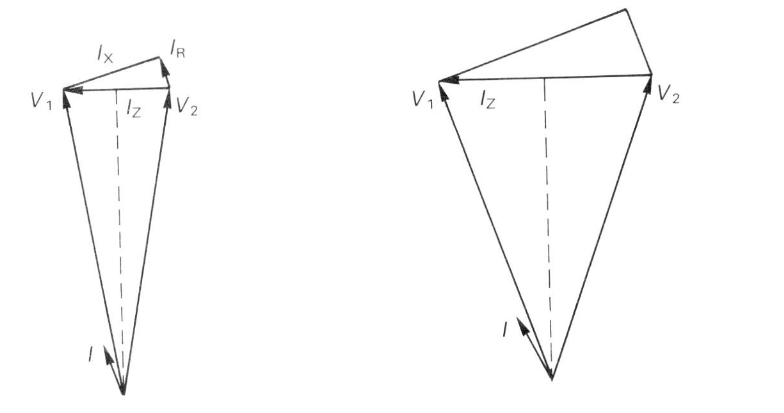

Figure 2.4 *Circulating load constraint for equal sending and receiving voltages*

factor reduces and the amount of MVAR which must be circulated increases faster than the MW transmitted, and there is a finite limit on the magnitude of MW which can be carried by a line of specific impedance. While this operating condition could be considered acceptable under emergency conditions, it is usual to provide some control procedure to permit a wider range of operable loads, and this involves a relaxation of the fixed-voltage constraints.

A further complication arises when a satellite load is supplied through impedances from two generation systems (Figure 2.5), as consideration has to be given to keeping this L_3 load voltage within acceptable limits for all conditions of load flows.

Voltage tolerance

It has been assumed that both generation busbars are operated at constant voltage, whereas in practice, automatic voltage regulators are only required to keep the voltage within a reasonable percentage of the nominal. A tolerance value of ±2½ per cent is quite usual for normal operating conditions, although a closer tolerance could be met. Thus at any time the two voltages could differ by up to 5 per cent without AVR action being implemented. Since the plant operated from the busbar voltages will function satisfactorily with a voltage of ±5 per cent from the nominal, no problems will be experienced.

Practical use can be made of this condition by setting the nominal busbar voltages at, say, +2½ per cent and –2½ per cent respectively,

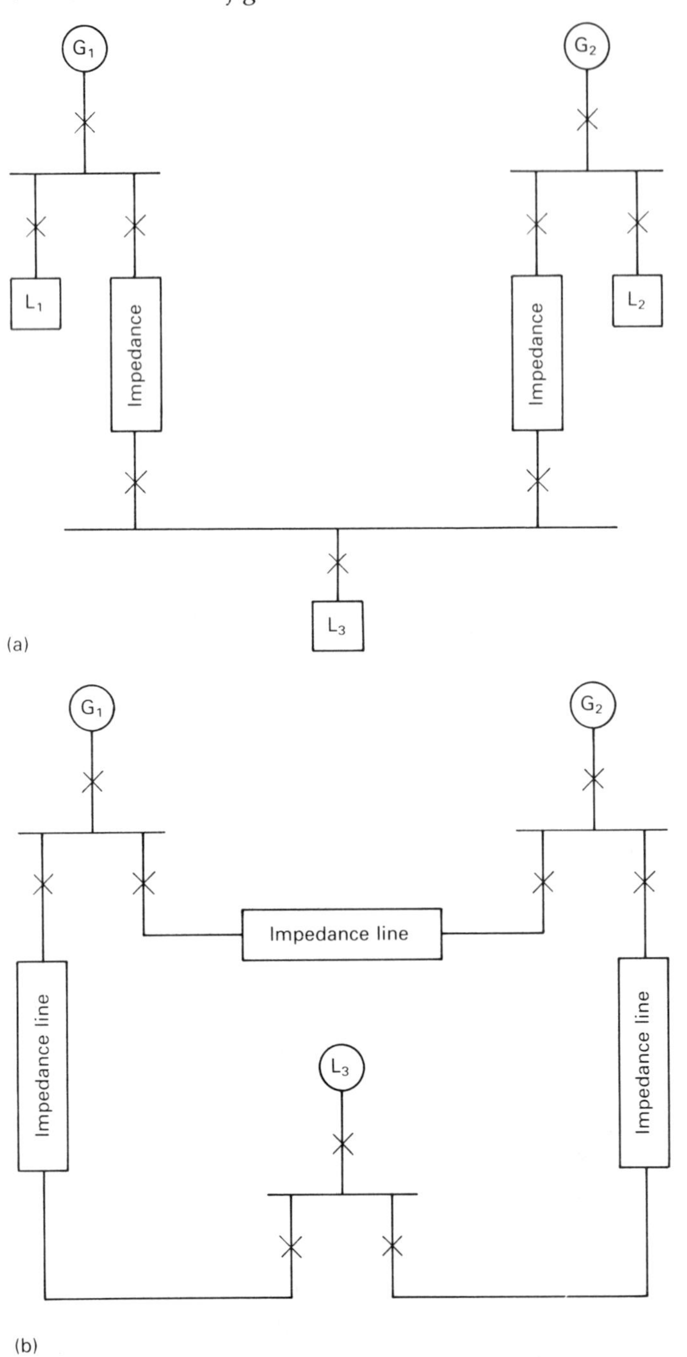

Figure 2.5 *(a) Impedance connected generators with intermediate load. (b) Ring impedance connected generators with intermediate load*

and using AVR accuracy of ±1 per cent; thus the voltages would differ by between 3 and 5 per cent. This could enable a reasonable circulation of power to be permitted as an emergency condition. However, if sustained operation is desired in such a mode, a more flexible form of voltage adjustment should be provided.

Voltage regulator

A simple autotransformer with on-load tap-changing can be used to inject a voltage difference between the two busbar voltages and so compensate for part of the impedance drop between the systems. This would require tapping steps of no greater than 2½ per cent, and with normal AVRs should give a reasonable control of VAR circulation.

When the distance between the two systems is considerable or the load to be transmitted is high, it will be necessary to use a distribution voltage greater than the busbar voltage. This will require transformers at both ends of the line, and by selecting suitable tapping steps on these a convenient control function can be obtained. With this arrangement, of course, there is no need for the two busbar voltages to be equal or even of the same order.

Modified AVR control

The MW load interchanged between the two systems depends on the magnitudes of the two absorbed loads and the governor controls provided for the two sets of generators. The procedures for MW control have already been discussed: it is only essential for the two systems to be compatible under steady-state and transient conditions to ensure stable MW control, the accuracy of which will depend on the sensitivity of the governing system and the mode of control adopted.

As the MW interchange load varies, so the interchange MVAR must vary to meet the voltage difference criteria, but the demand for MVAR increases faster than the corresponding MW demand. Thus if the receiving system could provide surplus MVAR it would reduce the line drop and reduce the MVAR demand on the sending-end system. Most modern AVRs can be adapted to operate as a constant MVAR or constant PF regulator for just such an operating mode. They function primarily as a fast voltage-regulating device, but with a slower control to adjust the excitation to meet the MVAR requirement, depending on the MW load if PF control is used. This adjustment results in a secondary change in the generated busbar voltage, but this can be limited to any required value, thus giving a cut-off to the amount of MVAR generated. Such a device has an automatic feature to restore the control function to normal AVR control, should the system revert to island mode of operation.

Figure 2.6 indicates what the value of MVAR circulated between the two generating locations, each with constant load, will be when the distribution of generators is changed, assuming that they are under proportional MW load share. The vertical axis represents the total field MW as a percentage of the total generation available; the horizontal axis represents the magnitude and direction of MVAR flow on the interconnector. Alternative characteristics are shown for the follower station operating at a constant PF of 0.85, 0.8 and 0.7,

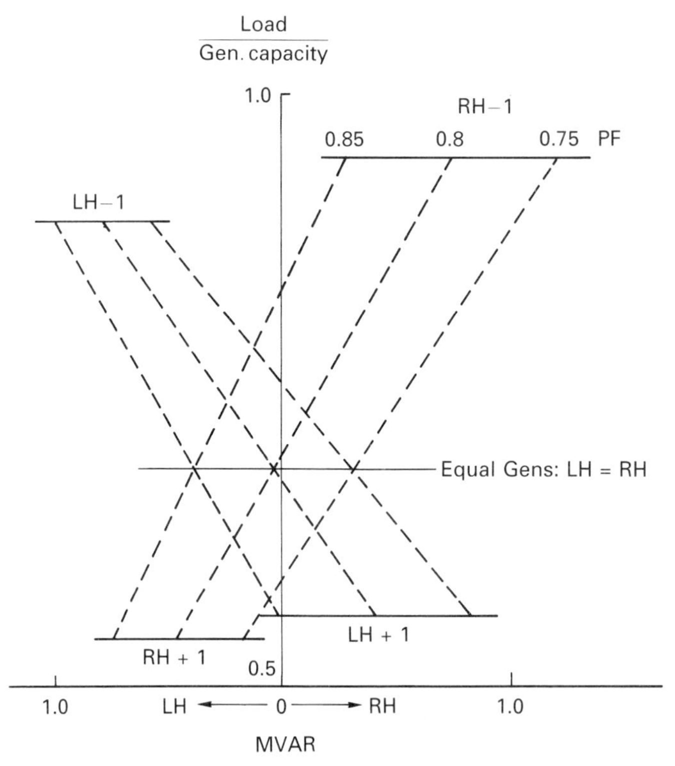

Figure 2.6 *Effect of constant power factor at one generator location (RH)*

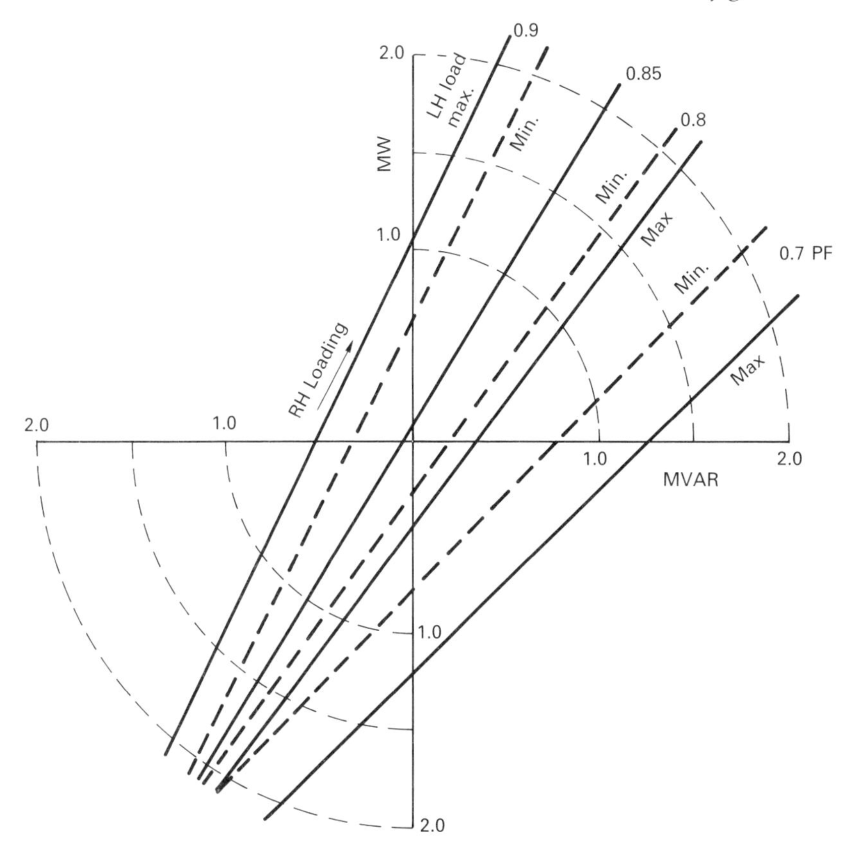

Figure 2.7 *Effect of load variation on interconnector loading*

and also for equal numbers of generators in each station, and for those being increased or decreased by one generator respectively.

Figure 2.7 indicates the interconnector load in ±MW and MVAR for different loads in the two stations for constant PF control at 0.8, 0.85, 0.8 and 0.7, assuming that the load PF is 0.85.

Mode of operation

When generators have a reasonable MW load, usually less than rated load to allow for system demand fluctuations, the generator heating is determined by the MVA loading and hence it can provide more MVAR than the rated value without affecting the life of the set. Such operation will also be within the rated capability of the excitation system and so, by more effectively utilizing the generation plant available, the loading on the interconnecting line can be significantly reduced.

The choice between constant MVAR control or constant PF should be made on the basis of normal variations of the system demand. The PF control will provide MVAR proportional to MW, but with constant MVAR control better line compensation is obtained for all

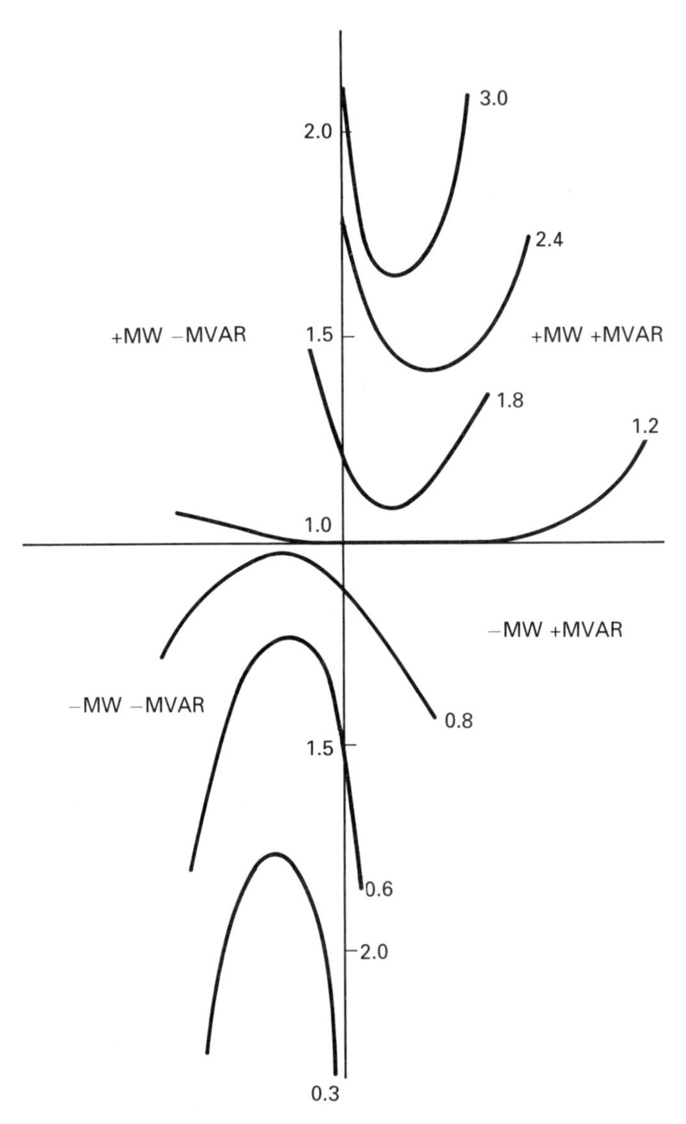

Figure 2.8 *Relative distribution losses*

loads. However, the value to be selected must be determined by the expected system island load to ensure that a reasonable surplus is available for line compensation.

When MVAR sharing between generators on a common busbar is provided by means of an astatic loop, this need not affect the above arrangements. However, if droop compensation is used it is desirable that both systems use identical controls or their difference may result in secondary MVAR flow in the line interconnecting them.

When the system is as in Figure 2.5(a) the simple control mode of using constant PF control on one set of generators is no longer adequate, as it is necessary to provide for voltage control at L_3 and this places a constraint on the simple control. However, if the two generator stations are interconnected, as in Figure 2.5(b), to form a ring system, the use of constant PF generation at one station is possible but may not give optimum operating conditions, as there are now two possible paths for circulating MVAR. With this configuration it is easy to control the MVAR circulating between the two generating stations when both are operating at constant

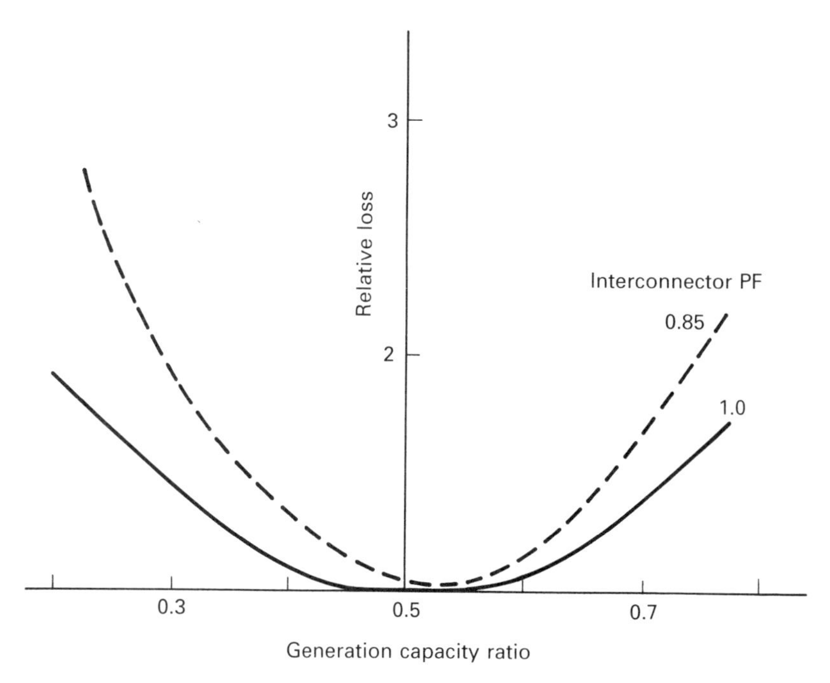

Figure 2.9 *Relative distribution losses for constant interconnector power factor*

voltage, and this then determines the distribution throughout the ring. The criterion to be adopted for this control can be based solely on minimum total distribution losses. Figure 2.8 gives a typical representation of the relative losses for the four quadrant values of the generation interconnector loading. Figure 2.9 can be derived and indicates the relative losses if the interconnector PF were controlled to 1.0 and 0.85 for alternative ratios of generation capacity at each generation station. Such characteristics apply to specific loading conditions but indicate practical control criteria for this simple configuration. However, even when using such control it is necessary to apply limit controls to ensure that the loadings of all plant and interconnectors are kept within safe design limits.

Other control criteria can be applied of a simpler nature; for example, the loading of the interconnectors can be equalized to ensure comparative useful life of the distribution components and this merely requires a comparison of the interconnector MVA or current when different operating voltages are used, in relation to their designed rating under their specific operating conditions in service.

Multiple interconnected island systems

The control procedure for two interconnected systems can be seen to be complicated but amenable to simple analysis. When more than two generation systems of comparable orders of magnitude are interconnected, although the principles involved are exactly the same, the multiple interaction makes the analysis of control much more complicated. While simple control procedures can be adopted which could maintain the system in steady-state stability, its behaviour under transient conditions, such as sudden load changes or fault conditions, cannot be simply predicted (Figure 2.10).

It is necessary, therefore, to adopt a control logic involving minimum disturbance but within as wide limits of response as possible, and by reiteration improve the accuracy of all the consequent function values to restore the ideal stable condition. This requires a determination of the most critical parameters in the system and the consequent results of variation of each of these in turn.

In addition it is necessary to evaluate the consequences of all abnormal loadings and plant and system faults, to determine

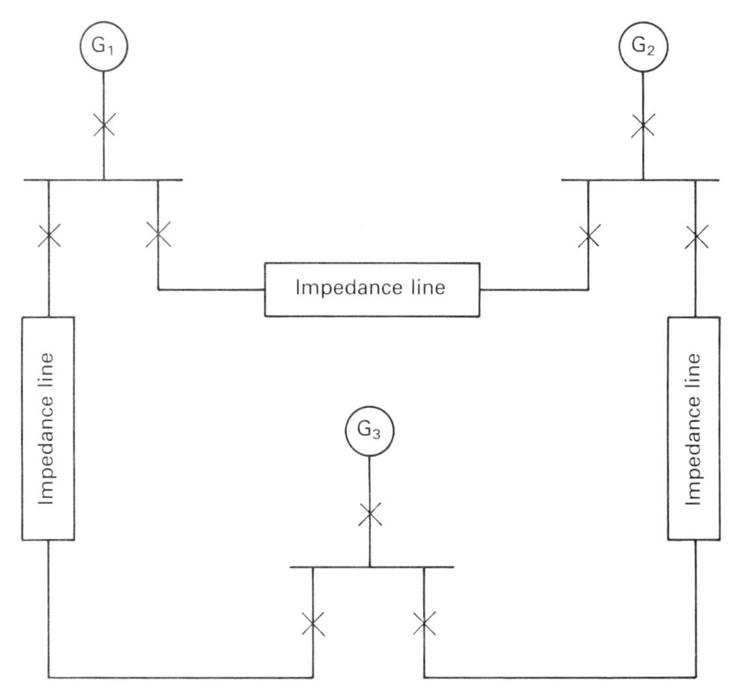

Figure 2.10 *Impedance connected generators in closed ring configuration*

relative dependencies of the residual healthy plant and load and identify the emergency actions that must be performed to maintain stability. This involves close coordination of the monitoring system, normally regarded as part of the protection system, and the emergency actions required to be implemented by the control systems, and these systems must be regarded and treated as a single control system involving the complete generation, distribution and load network.

Obviously, such a control problem is beyond the capability of a human operator because of the amount of data to be assessed, deductions of consequences to be made, and necessary actions to be implemented, all within the time when the complete system could lose stability. Consequently, while the steady control logic and control can be left to manual control, there must be some overall logic monitoring to assess safety levels of actions taken by the operator and also an overall protective system which would implement necessary emergency actions in the event of any serious abnormality.

This information can only be determined following a thorough assessment of the system behaviour under all operating modes and a competent safety audit of all the equipment and control logic involved, to ensure that the final system will fulfil its required duties in a safe and efficient manner.

To control such a system a PMS in some form will be required. The design and coordination of this is considered in detail later.

Satellite island system

A self-contained island system, one consisting of local generation and loads, may be connected to a large generation system (Figure 2.11) for a variety of reasons, such as the following:

1 Should the local generation be inadequate to meet the local load requirements, either due to increase of load or the loss or non-availability of local generation, the necessary make-up power can be obtained from the external system, thus avoiding the need to shut down any local loads.
2 When the local load includes plant which must be supplied continuously with power to avoid severe economic loss, a high-integrity continuous power supply can be provided by providing adequate local generation and also an interconnector of sufficient capacity to a large external system, and arranging for autotransfer from one to the other in the event of failure of either, while maintaining transient stability on the island system.
3 Where considerable local generation surplus is available, financial gains may be possible by running the system at rated output and raising revenue from power exported to the large system. This is particularly relevant for CHP schemes, where the waste heat from the generation sets is used locally in addition to their electrical output.
4 When the local island load is likely to develop rapidly, it is convenient to import power for a time to make up any local generation deficiency until the installation of additional local generator sets is economically justified.

With this satellite island system mode of operation, where a local island system is interconnected with a much larger system through an impedance circuit, the conditions existing are similar to those for

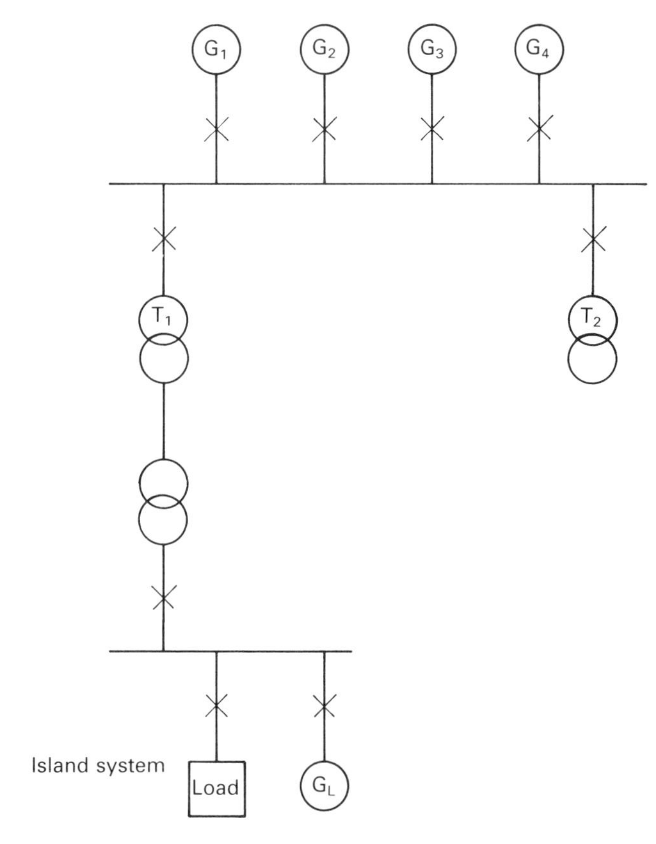

Figure 2.11 *Typical satellite connected island system*

two impedance-connected island systems. However, because of the relative size of two systems, the large system tends to behave as the theoretical infinite busbar system often referred to in textbooks, which operates at a constant frequency and generation voltage independent of the normal behaviour of the island system. As a result, the island system operates at virtually constant frequency and at a voltage determined by the interaction between the local generation excitation circuit and any voltage-regulation equipment operating on the interconnector circuit.

3

Generator application and control

The output of a single generator may be utilized in several alternative ways, and in each the mode of control can also be varied. The choice of application and control will depend largely upon the generator configuration being used. Since most generators will be required to operate in a variety of configurations, the optimum modes of application and control should be adopted for each, if practicable. Sometimes the alternative modes of operation will prevent the use of an optimum choice for each and it is then necessary to establish priorities and provide the optimum mode of operation for the most important or basic operating mode.

The choice of mode of application and control should be made before the types of plant to be used are selected, for the reasons explained in previous chapters, as certain specific designs of equipment are particularly suited to one mode of operation and may have characteristics which are undesirable under other modes of operation. However, in some instances it may be desirable to provide alternative generation units, each selected for its particular duty: for example, standby plant or peak-lopping sets which may not be required to operate continuously require characteristics quite different from those essential to base-load generation sets.

Normal and abnormal operational modes

When the system loading has been determined, using the criteria discussed previously, it is necessary to determine what conditions are to be regarded as normal, that is to say will be catered for by the generation provided without qualification. All other conditions will be classed as abnormal loading conditions and, for economic reasons, it may be desirable to qualify the performance of the generation system under some of these modes, or in extreme cases to exclude them from the range of practical operating conditions.

In some cases it is convenient to deal with some abnormal conditions in the same manner as fault conditions, by utilizing some ancillary sensing system to take alleviating action to enable the generation available to deal with the more important load units, e.g. by load shedding. This, of course, is only possible where significant load blocks can be given a lower operating priority than those regarded as priority loads.

An alternative procedure may be adopted where the electrical load can be regarded as requiring only low-grade performance: for example, some pumping installations can give a satisfactory output even with the system frequency reduced to enable the generation to deal with the consequent reduced load.

It is often difficult to differentiate between abnormal load conditions and fault conditions, because in many instances the consequential behaviour of the system is similar. It is customary to include fault conditions in any detailed study of abnormal conditions on a generation system.

Normal operational modes on small systems

On a small system it is practicable to determine the actual loading requirements with a reasonable degree of accuracy, including variation day to night and summer to winter, or on other regular shift-working bases, to give a steady loading pattern. Assuming that the variation of loading is slow and that no large loads are switched on or off suddenly, it is possible to select a number and size of generators to meet all conditions with reasonable efficiency. Depending on the nature of the load variation, it may be desirable to run all generators at equal percentage loading or, alternatively, to use some generators as fixed base-load machines, giving high operational efficiency, and use other sets for peak-lopping as required. This latter arrangement may have advantages if prime mover heat recovery is being utilized on the base sets.

Some small systems will have unique criteria to meet special load requirements, such as variable speed–frequency for marine propulsion, pulsing loads (either random or periodic), or very close frequency control for transmission equipment. These still enable a precise loading to be determined, and the application and control problems are quite explicit. Other conditions which require defining include high harmonic content or phase imbalance of loads, and these are matched to the capability of the plant design adopted.

When a small system is interconnected with a large system it is no longer necessary for the 'small' generation capacity to match the full range of steady loads, as import/export can be used to make up any differences, as required.

Abnormal operational modes

When a generation system has been designed to meet specific steady-state conditions it is necessary to consider its behaviour under abnormal conditions to determine whether it will continue to function satisfactorily, or to function with reduced performance, or fail to function. It may then be considered necessary to modify the design to upgrade the abnormal performance when it is required to recover from some specific conditions, or to give a better performance when they occur.

Such abnormal operational modes arise when the generation, distribution or loading system does not behave in the manner assumed when evaluating the steady operational modes. This can occur for a variety of reasons:

1 System loading being increased above design value.
2 Generation capacity being reduced below design value.
3 Change in system configuration and distribution.
4 Maloperation of control functions.
5 Major disturbances on the system.

These can be the consequences of:

(a) Incorrect action by operators.
(b) Incorrect action by non-compatible control systems.
(c) Failure of components in plant, distribution, generation or control.
(d) Electrical system faults.

Items 1 2 and 3 and (a) and (b) should be dealt with at the system/ generation design stage. By providing suitable monitoring, control and protection systems it should be possible to provide a system giving stability under these conditions and satisfactory system operation.

The failure of components can be considered on a statistical basis in conjunction with the choice of reliability level of equipment installed. Redundancy is usually provided to cover for failure of plant likely to impair safe and satisfactory system operation. This can sometimes be combined with other features such as load shedding and maintenance shutdown procedures to reduce the amount of redundant plant required for the abnormal conditions that must be covered.

System faults and disturbances can be detected and corrective action initiated rapidly by providing suitable sensitive and fast-sensing protective equipment. However, the action should not be limited to the distribution switchgear but should be coordinated with the generation control, including voltage and governor control, and this is one reason for designing an integrated control system for all aspects of control involved.

The analysis of such systems is, of course, complicated and involves considerable study before suitable designs can be determined. It requires very accurate data on all the items of plant involved as well as their control systems, and also accurate modelling of their behaviour under all ranges of variables resulting from the disturbances. This is considered in more detail in Appendix 3.

Generation control logic

Island mode – single generator sets

Simple manual presetting

When the characteristics of the electrical load are suitable it is possible to obtain reasonable automatic operation, which will regulate the generator parameters within their design range for normal loads within its rated capacity by presetting the no-load speed of the governing system, and the droop if used, and also the no-load voltage and voltage droop settings of the AVR–excitation system to suitable values. In applications requiring considerable frequency change, a volts-per-cycle regulator is usually more suitable than a simple voltage regulator. Many electrical loads reduce their MW demand as the supply frequency drops. This tends to stabilize the system frequency, as a droop-governed generating set reduces frequency as the MW demand increases.

The inherent voltage regulation of a generator results in the system voltage being reduced as the MVA demand increases. Thus if the system power factor remains constant there is a tendency to produce constant volts per cycle for varying loads on the system, which is ideal for induction-motor-type loads.

System loads may have other load-dependent characteristics and this simple control logic will no longer be applicable. When the system load can exceed the generator capacity, it is possible to add a simple load-shed system which senses excess load and arranges to disconnect non-essential units to keep the residual load within the generator capacity.

This simple control logic is ideal for low-grade power systems, such as some rural applications or isolated locations where service and maintenance cannot be guaranteed. The control devices used can be made quite simple and robust and the logic used needs little or no operator technical knowledge.

Manually adjustable control

When simple manual presetting does not result in the control system keeping the system parameters within a sufficiently close operating range for all loads, the same controls can be provided for adjustment by the manual operator for each significant change in load, to restore the values of frequency and voltage within the required range.

This system can be useful when load changes are infrequent and occur when the operator is available, but it should not be relied on when large load changes occur suddenly. The consequences of significant changes in system frequency and voltage can have adverse effects on residual loads on the system.

Simple automatic control

The above system of manual control can equally well be performed by an automatic control system which allows the inherent governor and AVR characteristics to respond to the load change and then slowly adjusts the set points to restore the system to the required operating parameters. This system is commonly used when load changes are relatively small or where the large excursion of system parameters following large load changes is acceptable to the system.

High accuracy control

The above control schemes are unsuitable where it is necessary to keep the system frequency and voltage within close tolerances for all steady-state loadings, or it is desired to give a reasonable range of parameters for specified abnormal conditions such as starting large motor loads, or recovering from particular load disturbances such as a pulse load.

Such control systems require special governor/fuel control equipment to give a high fuel ramp rate to assist rapid speed recovery, and possibly the addition of high mechanical inertia to minimize the speed drop resulting from a pulse load. Under such operating conditions it is essential to check the values of transient torques produced at all parts of the shaft system, as these can be several times as great as the steady-state values and may be oscillatory. It is also important to avoid possible resonances in the major components of the generator and prime mover comprising the set, as well as in associated components.

It is also necessary to ensure that the generator AVR–excitation system is adequate to meet the voltage conditions encountered under these transient loading conditions and maintain the system voltage within the required range.

Island mode – multiple generator sets

Simple manual presetting

The principles described for single sets can also be used with multiple sets, but even with duplicate generating sets the characteristics obtained from them will not be identical and may not be sufficiently closely matched to result in satisfactory operation under all load conditions. When the total system load is appreciably less than the total generator capacity available, the mismatch between set loading can be such as to leave the most highly rated machine within its capacity: satisfactory operation can be expected under steady load conditions. However, when the load changes are relatively large or fast, maloperation may result and loss of one or more set can occur. This becomes more probable as the margin between system load and generator total capacity is reduced or as the mismatch between generator characteristics is increased.

When the generating sets are not duplicates the above conditions will normally be more severe and it is unlikely that satisfactory operation will be obtained using a simple, manually preset system.

Manually adjustable control

This can be used for the above conditions where preset values will not give satisfactory operation. However, the operator must be aware of the relative magnitude and speed of the effects of all the control adjustments on all the sets involved, as incorrect sequence of control actions can aggravate system conditions and could result in loss of a set or, in extreme cases, in loss of the complete system, depending on characteristics of the system load transient behaviour.

Simple automatic control

Such a system will give safer control than the above and can be used satisfactorily for systems where the load changes are not rapid and are not of great relative magnitude.

High-accuracy control

Such a control system can give satisfactory performance on multiple generator systems for all load change conditions, provided that the dynamic characteristics of the generator sets, the generator excitation system and the prime mover fuel systems etc. are compatible and can be matched to each other to ensure satisfactory simultaneous transient responses.

Compound control

When it is desired to utilize some sets in different operating modes, i.e. all sets will not share MW and MVAR proportionally with their rating, additional control may be required because some sets will be operating on different dynamic characteristics as determined by their mode of loading, and there will be interaction between these sets during transient conditions which must be within the control capabilities of these sets.

Island interconnected to large system

This condition is much more complicated than simple island mode operation and involves several factors in the energy/economics operation, many of which may be indeterminate or may vary with time.

Governor control

In this operating mode the function of the island governors is quite different, as the large system will maintain the system frequency within close, possibly statutory, limits, since its relatively high inertia and generating capacity will mean that load fluctuations on the island system will not have a significant overall effect on the total system. Such island changes in load will, however, result in variation in the power interchange between the island and the main system, and the main function of the island governing system will be to determine the magnitude of such fluctuation.

Adjustment of the governor steady-state setting will determine the island MW generated, and the difference between this and the island MW load will determine the balance to be made up as input from, or export to, the main system. Obviously the power tariff will determine the most economic setting for the generated power or control mode to adjust this with island load variation.

When waste heat recovery is being used on the generator prime mover, the economic value of this must also be incorporated into the energy cost equations, as it is actually preferable to utilize this up to the set's rated capability when this is the design point – but this is not always true. When more than one set is available, mixed operation can be adapted, with the heat recovery unit operating as a base load set at optimum operation and the other as a varying load machine used to control input/export and not using waste heat recovery. Other factors can be involved, such as maximum set operating efficiency, reduced maintenance requirements due to elimination of load cycling, variation in utilization efficiency, or demand for the waste heat.

The number of permutations possible can be quite high, particularly if there are several island generator sets. The energy equations must be evaluated in relation to operating costs before the best operating modes can be determined for possible alternative loading conditions. These factors must be assessed before the simplest electrical control procedure can be determined to suit the electrical system requirements.

Excitation control

Variation in import/export power will produce a variation in voltage drop in the interconnecting circuit impedance and this will affect the island voltage and result in MVAR variations. The island generators must be capable of operating under such conditions, and the AVR–excitation system must be designed to keep the values within acceptable limits for the plant as well as the generators.

As previously described, use of a simple voltage control system on the island generators would determine the MVAR import/export in relation to MW and these might not be acceptable for technical or economic reasons. It could result in unnecessarily large fluctuations in island generator MVAR loadings or in the transmission system.

In this operating mode the generator excitation control should be used to keep the MVAR flow within specified values, either as determined by the island generator capacity or the permissible interconnector loading capacity (MW plus MVAR), actual, or as determined by economic factors.

The actual excitation control procedure should be selected on the basis of economic factors as determined by the import/export power tariff, and particularly by the significance of PF or MVAR for the costs charged. When no commercial benefit is to be gained from exporting lagging MVAR to the main system using the established tariff, it is preferable to minimize or eliminate the export of lagging MVAR, or even to import lagging MVAR from the system if there is no significant penalty. This can result in higher island generator efficiency, reduced operating costs and extended useful operating life of the plant.

However, if the financial return for exporting lagging MVAR to the system is sufficient, the island generators can be controlled to give maximum lagging MVAR up to the full MVA rating of the sets and this could supplement the remuneration obtained from exporting MW to the system.

Similar conditions can apply when importing MW from the system, as the island generators can still be operated to produce maximum lagging MVAR if financial gain can accrue from reducing that imported from the system.

Since there is a direct interaction between the island governor controls adjusting MW flow and MVAR flow, the excitation system must be compatible with regard to rate of response and consequent range of control of MVAR from the generator, and those of different designs of generation sets must also be compatible as the integrated values of MW and MVAR of all generators determine the import/export values at any time.

4
System control logic

The prime requirements of an electrical load are the available voltage and frequency. These are the primary factors in all electrical system control and determine the resulting MW and MVAR power flows. However, when the load and generator terminal conditions differ as the result of an impedance of some form being connected between them, the load and generator voltages will no longer be equal and the control system will require additional parameters to be involved in the control logic. The interaction of all these functions is such that it is customary to arrange control systems to maintain the significant parameters within acceptable limits rather than attempting to maintain these at fixed values; this greatly simplifies the control procedures.

These supplementary functions are usually limited to two of the following: MW, MVAR and PF. That is, we are still considering only an electrical system. As mentioned in Chapter 3, a result of utilizing waste heat recovery on the generating sets is that the logical implications of control are modified: these now involve the new concept of adding energy to the control logic; and thus considerations of economics and finance.

Power management systems

As a result of this escalation in electrical system control parameters there has been a tendency to utilize power management systems to help the operators to maintain and control the system. A PMS may provide the necessary data in a convenient form to enable the operator to make the necessary control decisions; alternatively, it may include logic to analyse the data and provide recommendations as to the necessary or desirable control actions. In some cases, such a system will initiate the control functions without operator intervention. Such control can be limited to the basic parameters on the available generating sets, or can be extended to a wide range of other operational activities if required. It may also be extended to other services such as monitoring, data collection and processing, prediction and dissemination, as required. PMSs are normally processor based and can interface with a wide range of associated equipment, including standard conventional communication systems.

The term PMS was originally used to describe equipment supplied with prime-mover-driven generator sets which arranged for the automatic run-up and stopping of the set. By measuring the load demand and the available power from the connected generators, the system could automatically ensure that sufficient generation capacity was available to meet the current load demand. The primary purpose of these first units was to give an automatic start of the prime mover and, by including an autosynchronizer, to automatically put the generator on the bars. It was a simple extension to include governor control, arranged to apply MW load to the set when started and synchronized, and such control was also utilized to arrange proportional sharing of the MW load between all the generators on the bars at any time. As the PMS controls the set governors, it can also be used to give the required mode of frequency control, either droop or isochronous.

The function of such PMSs was purely mechanical in that they were used to monitor and control power in the form of MWs generated, but their usefulness was soon extended. From the data made available to it for MW control, the system could also be used to monitor and, if necessary, control MVAR sharing between the generating sets. It also made simple the use of mixed operation modes using selected generators for base-load operation, or adjusting the proportions of MW and MVAR on generators to suit their optimum operating characteristics. Having MVAR monitoring and control available, it was easy to incorporate motor starting supervision which determined if there was sufficient MVAR available to start any large motor on the system and, if not, warned the operator and locked out the motor starter until satisfactory conditions were established, e.g. by the PMS starting additional generator sets.

As the usefulness of the PMS became appreciated, and experience proved the reliability of properly designed systems, other features were added to deal with the system rather than to control only the generator sets.

Load-shed systems could be included by providing the PMS with data on loads which could be expended in an emergency, such as system overload with inadequate generation or loss of a generator set due to a fault or to system disturbance. Data provided included status, i.e. running or not, loading, either MW or current (when a typical operating power factor would be assumed), and the order of sequence in which selected drives should be shed. A simple addition to this system was to provide an automatic restart (reacceleration) facility which could be used to restart those drives that had been load-shed when normal, healthy conditions had been restored to the system.

Some systems are designed for alternative modes of operation. For example, part of the plant, regarded as essential, is maintained with adequate generation at the expense of other sections; a suitable PMS, which monitors the status of all the relevant system switches or circuit breakers, can control the total system normally or, under specific conditions, can split the system and provide suitable forms of control for the two, or more, subsystems. This logic can be extended to interconnected generation stations which can each be provided with island mode control, but in which the interchange of MW and MVAR can be supervised by the same PMS with optional modes of operation.

Other specific control functions can easily be added to such PMSs, since the essential input data are already being provided and only supplementary logic is usually required; this can enable a fully interconnected system to be controlled accurately and safely.

Many supplementary features can be added to the relatively simple basic control functions to deal with practical operating difficulties and to extend the stable operating capabilities of the total system. They can be interfaced with the protection system to render it more effective and to improve system stability. These are more often associated with interconnected systems which can present severe operating problems if left to manual operation.

Island systems

In an island system the total MW and MVAR demand is determined by the connected load and the system voltage and frequency, and the total generating load must match these values. The behaviour of electrical loads has been examined in detail together with their significance with regard to generator behaviour, and the necessary governor actions to deal with MW loading (and frequency) and excitation control to deal with MVAR loading (and voltage). To provide a stable system with closely controlled voltage and frequency it is necessary to ensure that adequate prime mover power is available to meet all steady MW demands, plus any MW demands due to transient conditions such as accelerating motors and loads, and that adequate generator excitation is available to meet all steady MVAR demands, plus any MVAR demands due to transient conditions such as starting and accelerating motors and loads. By measuring the steady MW and MVAR load demand and relating this to the available generating capacity it is possible to determine the available surplus. In systems where the load changes are gradual it is possible to determine the point at which the load must be reduced (by load shedding) to meet the capability of the existing generation or, alternatively, at which additional generation capacity must be added to the system. It is, of course, necessary to ensure that the rate of load growth is not so great as to exceed the generation capacity significantly before the rectifying action has been taken. Any excess would result in reduction of system frequency or voltage or both, and while with some types of system load this could result in a stable system, with some particular applications instability could result. In addition to assessing the steady load margins, however, it is necessary to determine what transient loads have to be provided for. These could be in the form of large steady loads being suddenly switched on, or dynamic loads, such as the starting of large motors, involving accelerating loads as well as final steady-state conditions. Another condition frequently considered in this category is the loss of a generator set due to some fault in prime mover or generator. These conditions could result in system instability before load shedding or additional generation can be effectively implemented.

Two alternative approaches are available for additional loads: (1) allow a running MW and MVAR margin in excess of the greatest single load likely to be started; (2) inhibit such starting until adequate margins are available. With the former system there remains the possibility of starting two large loads within a short space of time; this condition is prevented if system (2) is used. Using the lock-out system might prove restrictive to production if additional generation could not be provided quickly, but this could be offset by load shedding if there existed an adequate amount of expendable, non-essential load.

The loss of a generator, which can be instantaneous, can only be dealt with by operating with full redundancy, i.e. where the margin between generating capacity and running load is greater than the rating of the largest single generator set. Such full redundancy often solves the transient load problem, since the greatest single load is actually less than the rating of the largest generator. PMS functions required to deal with the above conditions involve generator set start–stop, load shedding and drive lock-outs.

Generator set start–stop

Starting a generator set involves several distinct procedures: monitoring all conditions healthy, starting and accelerating the prime mover, exciting the generator and synchronizing with the supply, and loading the set to the required MW and MVAR values. Each of these functions can be simple or complex depending upon the size and type of generator set. In some instances it may be advantageous, for economic or safety reasons, to carry out some of them manually before allowing the PMS to complete the operation automatically. For example, to provide a fuel system under PMS control to check availability and quality of fuel, correct supply conditions etc. could involve considerable hazard and would require equipment of very high reliability; it might be simpler for the operator to check and set up the fuel supply to the prime mover before handing over the autostart control to the PMS. This also applies to electrical switchgear where the operator must have the equipment in an operational mode: it is then only necessary for the PMS to check that this condition has been met.

Most prime movers provided for generators have an automatic start system with inbuilt checks to ensure that safe conditions are being met and operations such as purging or load applications are correctly carried out. Such routines are ideal for inclusion in the PMS overall sequence control, but it is still necessary to ensure that all the signals needed to indicate healthy conditions are available for the PMS to check that the start sequence is proceeding correctly.

The PMS, having been advised by the operator that the generator set is healthy and available, can then initiate the prime mover start, check that run-up has been correctly completed to synchronizing speed, excite the generator, and arm the autosynchronizer and allow it to close the generator onto the bars. It will then adjust the generator MW and MVAR as required by the load conditions.

Normal shutdown would be carried out in the reverse sequence by unloading the generator, opening the generator breaker, and then decelerating the prime mover according to its inbuilt shutdown programme, leaving it in a condition ready for a restart if required.

It is quite usual to permit the operator to carry out these separate stages of start or stop using the PMS control procedure, and this capability is particularly useful when commissioning or recommissioning a generator set.

When the PMS controls several sets, the desired selections sequence is determined by the operator to meet the criteria of balancing hours run, minimizing running of a defective set or optimizing higher-efficiency set running, or other conditions considered significant. Should a PMS start sequence on a particular set fail to be completed correctly, it can select the next set in the sequence and start it, while giving an alarm on the faulty set.

It is often beneficial to restart a set which is still warm after being recently shut down, and this could be arranged to override the next set in the start sequence if demanded within a specified time of its being shut down.

Starting and stopping sets imposes additional stresses, both thermal and mechanical, and these operations are kept to a minimum consistent with economic running. Thus, even when the system load has dropped to a level not requiring the next set on the stop sequence, there will be a period of delay before the stop is implemented: only if the load remains below this level for this complete period will the stop be implemented.

Another logical concept used for start–stop control is load prediction. By analysing load–time data it is frequently possible to predict whether a load change is likely to be maintained to a point requiring start–stop action or whether the change will reverse, requiring no action. Quite often, load–time profiles follow known contours and large-load switching may be carried out at regular intervals. Such information can be utilized by the PMS to determine prospective set requirements.

Another condition which can usefully be incorporated in the start–stop sequence is any indication of defect in one of the generator sets. An alarm indicating the possibility of a shutdown, either healthy or as a result of a developing fault, can be used to start the next available generator set on the start sequence. When this standby machine has been started and is carrying load, the faulty set can be unloaded; the operator then has time to assess its condition without anxiety about possible disturbance to the system if it were to shut down while carrying load required by the system.

Some prime movers require quite a long run-up, particularly when starting cold, while others are capable of starting very rapidly. Normal times can vary between 1 and 20 minutes. This must be borne in mind when finalizing start-up logic: a load prediction logic can be advantageous when the standby set has a long run-up time. Where mixed prime movers are installed, the run-up times should be considered when nominating the running sequence. As far as possible, the slower run-up sets should be utilized as running machines and the fast starters used for standby. The starting time can be a very significant factor when selecting a new prime mover for a generator set, as it can have a direct influence on the stability of the system.

Load sharing

When a generator set is started by the PMS, or manually by the operator who then transfers control to the PMS, that system will then control the generated MW and MVAR on the basis of some predetermined criteria. The simplest system is to load all available generators in proportion to their rated MW and MVAR, which implies that prime movers and generators run at about the same percentage of their capability and may therefore be expected to have similar prospective useful operating lives. However, there are conditions which could justify an alternative arrangement. The set efficiencies vary with load and this variation can be significantly different between different types of prime mover or between similar sets using different fuels. Where a significant cost saving could be obtained by running one or more sets at optimum efficiency, these sets could be operated as base-load sets running at fixed load as far as possible consistent with power demand requirements. Other sets could be operated at reduced loads, where the reduction in efficiency is more than offset by the increase in that of the base-load sets.

As the efficiency of the generator is very high compared to that of the prime mover, this logic does not apply to MVAR sharing; the base-load machines should be operated at about rated PF and the balance of MVAR made up on the more lightly loaded generators, which can share the residual MW and MVAR in proportion to their rating.

In practice, there is no great advantage in sharing loads accurately, as the sets are never operated close to their operating limit, except during an emergency such as loss of a generator. This eliminates the need for highly accurate measurement and reduces the number of operations required by governor or excitation control. Small changes in load will seldom require any control action if reasonable tolerances are allowed on the magnitude of MW and MVAR loads shared by the sets.

Changes in load will result in a frequency change and this can be allowed a fairly wide tolerance, again reducing the number of control actions carried out. Most island systems have a stable load–frequency characteristic and it is operationally desirable to let the governors function on a 'droop' mode and only to use frequency adjustment, which must be applied equally to all generators, as a slow compensation. This logic of governor control has distinct advantages when emergency or fault conditions arise which demand fast responses to assist system recovery.

The excitation system control can function in a similar manner by allowing sudden MVAR changes to be compensated for by voltage droop AVR control and to be associated with a slow-response voltage adjustment which will produce steady conditions after transient conditions have disappeared. Mismatch in individual generator set MVAR sharing due to differences in their impedance and saturation characteristics is not significant and the wide tolerance allowed can avoid unnecessary control action during such disturbances.

Load sensing

The logic used for load sharing can be extended to determine when set starting or stopping should be implemented or when load shedding should be initiated. The total load demand, MW and MVAR, is evaluated and related to the available generation capacity to determine the available margin and compare this with the margin considered necessary for satisfactory operation. This must be done on steady-state values for normal loading, but it is also necessary to determine the transient margins, as these may be limiting for some conditions.

The steady-state capability of a generator set is often taken as the set rating, but this is not necessarily always correct. Prime movers such as gas turbines are very dependent on the ambient air conditions and their power capability can reduce rapidly as ambient temperature increases. Likewise, blade condition can be very relevant: if significant fouling is present, this also will reduce output. It may be advantageous to relate output capability to inlet air temperature and set outlet temperature, but this, while more accurate, requires more input data due to the monitoring system together with specific design parameters for the engine itself.

Other prime movers are designed to be capable of a sustained overload for a specified time and this may eliminate the need to start additional generation for short-time increases in system load demand.

Other factors, such as fuel characteristic or available cooling water temperature, can affect output capability. When determining the set capability in MW, either safe estimates of all these factors have to be made, or the relevant parameters must be measured and supplied to the PMS together with the relevant performance equations or simulation.

The electrical rating of a generator is a nominal MVA value at a nominated PF, but the latter factor is related to excitation requirements and can provide a limit in its own right. However, it is frequently found that this has a higher inbuilt limit provided for other operational purposes and the operating limit may depend solely on MVA. This limit is temperature based and consequently depends on design conditions of cooling being provided if rated output is to be maintained. When the total generation output has been evaluated and related to the currently available set ratings, the difference gives a working margin instantaneously as a steady state, but the PMS can also evaluate the effective transient margin. The most common working margin is that corresponding to the output of the largest running set giving 100 per cent redundancy: that is, if this set were lost, then the remaining loads on the other generators would correspond to their full rated capability. While this gives a theoretically satisfactory condition, it is not always practical, since any subsequent load increases could jeopardize the generation system. It is customary to associate such operation with a load-shed system which would operate to ensure that the remaining sets were kept within their design rating. The advantage of the 100 per cent redundancy is that it usually ensures transient stability following loss of the generating set, which might not be obtained if the corresponding load reduction were to be provided solely by load shedding.

Obviously, the more important the system load, the greater is the emphasis put on continuity of power supply: the working power margin will correspond to full redundancy. However, with some types of load and on systems with mixed loads (some requiring a high level of continuity and others which can be considered as temporarily expendable), it is possible to operate with smaller power margins and allow some sections of load to be disconnected during times of severe disturbance. Suitable protection schemes can be adopted for this function.

Load shedding

It is quite usual with a PMS to monitor not only the individual generator MW and MVAR to determine total loading but also the MW and MVAR in feeders from the generation busbars. This gives a useful cross-check to validate the loading values used by the PMS to control the generator operation. It also provides data on the status and magnitude of the main generator loads and whether any of these can then be regarded as temporarily expendable; the PMS can then assess whether shedding such loads would result in a stable system. All expendable loads have to be identified to the PMS and given an order of priority. Thus, if a gradual increase in steady load reduced the generation power margin below the nominated level, the PMS could disconnect sufficient of the expendable loads to restore the required margin. If sufficient working margin of expendable load was not available, the PMS would warn the operator, who would then start another set – or the PMS could do this if provided with the start–stop facility. To minimize the amount of load shed the expendable loads could be tripped individually, at short time intervals, so that the monitoring system could stop the load shedding as soon as the working margin was restored.

However, there are other conditions which can be greatly assisted by a more rapid load shed. These include all transient effects due to system switching conditions or faults which limit the effectiveness of the generation system. Thus if a large block of load were to be switched onto the bars, significantly reducing the working margin, then the corresponding shedding of an equal block of load would restore the original situation. If the PMS is notified of the amount of load to be switched, the load-shed system can predict the number of expendable loads in the sequence which must be switched off to maintain the margin.

A corresponding action can be taken if a loaded generator set is tripped off the busbars and the PMS can predict the loads required to be shed as a block to restore the required margin. This is one facet of control in which load prediction can be used to estimate the prospective demand of the system, based on measured trends and a knowledge of typical trends and time patterns of the load, and prepare an adequate load shed if the actual load does materialize. The choice between starting additional generation to meet the prospective load demand, discussed previously, and load shedding may be determined by the actual time required to get a set started and loaded. A combination of the two systems could be a good compromise: the load shed would give immediate restoration of the working margin and cover the time required to start additional generation.

Depending upon the amount of reduction of the working margin, it might be considered an acceptable risk to dispense with such load shedding, but this can only be decided when details of the actual system involved are available for analysis. When a generator set is tripped off the bars when carrying load, the disturbances to the busbar voltage and frequency are relatively small. Provided that the system had 100 per cent redundancy, stable conditions will quickly be restored.

However, if the set is tripped off as the result of an electrical fault, the resulting disturbance may be considerable. The voltage at the fault would reduce to zero and large fault current would flow to this point, resulting in disturbance to the complete network voltage. The reduction of busbar voltage would result in the generators having a severe load reduction and the prime movers would start to overspeed, producing a rise in system frequency. As soon as the fault was isolated, the voltages would try to recover to normal and the restoration of the generator loads would result in a reduction in system frequency. During the reduced voltage conditions, induction motor drives would decelerate as a result of inadequate motor torque and restoration of voltage would counteract this effect. Whether the voltage recovery would be sufficient to reaccelerate the drives would depend on several parameters. Usually, a detailed transient analysis study is required to determine whether the system would recover stability.

To maintain stability it is essential to restore system voltage rapidly and also to provide enough power to maintain existing drive loads as well as providing power to reaccelerate. Under this condition the shedding of all non-essential load can have a significant effect in assisting system recovery. The transient load shedding differs from the steady-state load shed in that it is initiated by the protection system which detects the fault condition and it trips all available loads simultaneously, as rapidly as possible.

Reacceleration

When satisfactory system conditions have been re-established following a load shed, the operator can restart the drives that have been shut down, provided that adequate generation capacity is available. This function can also be performed by the PMS, which has all the data relevant to the drives that were running and that have been included in the load shed. The PMS must be provided with a restart sequence, which need not be the same as the load-shed sequence. Where the drive is part of a process and interactive logic is required to control the component drive starts, this too can be incorporated into the PMS if desired. Alternatively, the PMS can advise the process restart logic when conditions are reached which permit healthy restart. Conditions for a satisfactory restart include recovery of the system voltage to a reasonable value which will allow healthy motor start. Industrial drives are usually specified to be capable of starting with a voltage of at least 85 per cent of normal, but it is preferable, when restarting non-essential drives, to wait until a higher system voltage has been obtained (usually 95 per cent of normal), as this produces less adverse effects on the generators and gives a quicker drive run-up time with less stress on the motor windings.

Load control

When relatively large loads are connected to the system they can produce adverse effects on the generator sets and also on the other drives on the system. Large induction motors which are switched direct-on-line can demand as much as 10 times their normal MVAR rating, although the MW demand is usually less than their rated MW during starting. This large MVAR must be provided by the generator excitation system, which also has to maintain the system voltage. The excitation system is designed to maintain rated voltage over a specific range of MW and MVAR ranges, and any demand in excess of this will result in reduction of the system generated voltage. This will affect all the connected plant and will also reduce the starting capability of the motor being started. The severity of the voltage dip depends upon the generator impedance, the measure of excitation forcing available (ceiling voltage) and the magnitude of the MVAR load being demanded.

Such a large starting MVAR will have even more severe effects if there is an appreciable impedance between the generator and the motor, such as a step-down transformer, as the voltage drop in this impedance will be greatly increased and, when combined with any generator terminal voltage drop, will greatly reduce the motor torque available for starting and hence its ability to accelerate its load. In extreme cases the motor could fail to break away, or could break away but fail to run up to full speed, both conditions being dangerous for the motor and requiring protective devices to detect such a situation and disconnect the motor.

The PMS has the data available regarding the existing system MW and MVAR loading and can be provided with the generator capability curves which identify the range of MW and MVAR over which its excitation system can maintain rated voltage on the bars. If the PMS is provided with the starting characteristics of all large motors, or any other loads, it can determine whether such a load can be connected to the system without producing adverse operating effects. Before starting such a load, the request is first passed to the PMS, which checks the existing generator capability; if it is adequate, the PMS refers the start request signal to the appropriate circuit breaker and starts the load. However, if there is insufficient generator capacity, the PMS locks out the load and alerts the operator that it is unsafe to start it until the existing system load has been reduced or additional generation has been made available.

It is possible to use the PMS to carry out necessary load shedding or start additional generation to enable a particular load to be started, or alternatively to provide information for the operator as to which loads can be safely started with the current system load and generation capacity. This enables the operator to provide the necessary conditions before attempting to start a particular load.

In island mode, where usually only a few generators are connected at any one time, starting a large motor will have severe effects on the system and these should always be evaluated before a start is attempted. The PMS can be arranged to carry out this function for the operator. In other modes the consequences may be more significant with regard to the effects on local load, but not necessarily so on generation, although some particular generators may be adversely affected if they are closer to the load than other generators in the system.

Subsystem operation

Even on quite small systems it is usual to provide bus section switches which can separate the main generation busbars into at least two independent sections if required; when multiple generation sets are installed, these are distributed onto the separate sections of busbar. Thus it is possible for the loads connected to a particular bus section to be supplied by the generators connected to that same section, and to be electrically separate from the other system loads and generators which are connected to other busbar sections.

With the majority of systems this condition is regarded as being only an emergency condition and is not normally used. However, it can have significant advantages when maintenance work is being carried out or when modifications are to be made to the system. It is sometimes used as an operational condition when one set of loads has to be maintained at a fixed frequency but the other bus section, connected to its own generation, can operate at a different or variable frequency to suit the specific needs of part of the system load, e.g. by adopting variable-frequency starting to run up a very large motor which, when up to full speed, can be synchronized with the rest of the system by synchronizing across the open bus section switch.

Where island systems can be interconnected, the same conditions can arise when each island system can be regarded as a subsystem of the main interconnected system.

A PMS system can be used to provide any of the functions already described for the main system as a whole, but it can be organized to provide these same functions to each of the subsystems or to any groups of subsystems. The PMS monitoring system can detect from the status of the various bus section or interconnecting circuit breakers which plant, including loads and generators, is connected to each discrete subsystem, and can control this in the normal way. Recombination of subsections is automatically detected by the PMS, which automatically integrates the loads and generators into a new system configuration. It can also combine operation schedules, such as generator start sequences or load-shed sequences, to form new schedules for the reconfiguration, and these can be preselected by the operator or selected by the PMS on a nominated logic base, if required.

Some functions, such as load control or starting large motors, can be more limiting on a subsystem, where the connected generation available to provide the necessary MVAR is usually much lower. In such instances special precautions usually have to be taken, and the PMS can be organized to alert the operator to these.

When the main system is a distinct unit it is only necessary to use a single PMS unit which can control all possible subsystems. When the main system consists of two or more discrete subsystems it may be advantageous for each to have a PMS unit to permit separate subsystem operation, where all are interconnected and operate as slaves to a nominated master. This arrangement has advantages when the subsystems are physically separated by considerable distances. Since each PMS unit contains the full logic capability, one can act as master and the others as slaves, any of which can be arranged to take over the function of master if required.

Safety systems

Because of the information available to a PMS it is convenient for it to undertake functions which will assist in maintaining the power system under abnormal or adverse conditions, or, in the event of failing to maintain the full system, in shutdown to a safe condition. It can also include the capability of assisting or controlling a recovery to normal healthy system conditions. In this context, the term 'safety' can apply to the generation, distribution and electricity supply system, including the actual users, to the plant forming the system, to the operators of the plant or, in special cases, to the structures containing the generator plant or to personnel associated with it.

Continuity of supply involves generation and distribution. The maintenance of adequate MW and MVAR generation has already been discussed. Reliable distribution involves suitable switchgear with bus section isolators to enable disconnection of a faulty action to be performed while maintaining the system on the remainder of the busbars. Distribution transformers have to be adequately rated for their maximum loads. To increase integrity, multiple units are preferred, connected to different sections of the busbars in such a way that the loss of one transformer will not impair the supply to the load. Such transformers usually feed lower-voltage busbars which should have bus sections corresponding to the higher-voltage busbars. Thus the load on the lower voltage busbars is split between two or more transformers but, should one of these fail, the load on the corresponding bus section can be fed through the bus section switch (being closed), or a standby transformer can be switched in. A PMS which monitors the system configuration and measures all transformer loads can determine appropriate action in the event of a transformer fault occurring and can bring on the standby supply without the necessity to shut down and restart the load on the faulty transformer. Similar PMS action can be taken with other major items of plant which fail, provided that suitable redundancy is available.

When the plant outage is so severe that the PMS cannot recover the complete system load, it is possible that it can be arranged to segregate essential drives and arrange a supply to these from remaining generation or, in the extreme case, can be arranged to start emergency generation and connect the essential load to this. All this can be done automatically, if required, by the PMS.

It is sometimes necessary to shut down healthy generator sets or switchboards, not because of any electrical plant fault but because of a hazard produced by the process plant or load, or even by the prime mover fuel supply developing a fault. In extreme cases there could be serious flammable gas or liquid release, or a fire could break out.

Warning devices are provided to try to protect against such occurrences developing into disasters. Fire and gas detection systems are available as standard self-contained types of system. These are sometimes combined with emergency shutdown systems which integrate the alarm data and decide and initiate required actions. These units perform the same logic operations for which the PMS is designed and such actions could well be performed by the

PMS, but the latter has the advantage that it possesses much more information about the state of the electrical system, including generation, distribution configuration, and loading, and is capable of making more logical decisions as to the total effects of alternative actions which might be indicated by the fire and gas or emergency shutdown systems. Alternatively, the PMS can be operated in conjunction with standard fire and gas and emergency shutdown systems which will initiate necessary actions local to the relevant hazard and plant associated with it: the PMS can be arranged to take supplementary action on associated plant to maintain, as far as possible, healthy supply conditions to the system outside the hazard zone.

Following any shutdown of part of the system, the restoration of supply can be a long and complicated process, particularly if some equipment has been damaged or is in some way unavailable for the start-up procedure. It is possible to translate the safe restart sequence into a logic which the PMS can use. By monitoring the relevant functions, it can start generator sets, energize the system and prepare for the reconnection of its load when conditions are suitable. Alternatively, the PMS can use the restart logic to monitor the actions taken and correctly completed (visually and also as a permanent record, if desired) and give an alarm if the sequence is not being followed correctly or a wrong action is initiated. In some cases an operator can perform the converse operation, allowing the PMS to carry out the restart sequence as a series of discrete actions which can be monitored individually before the PMS is allowed to proceed to the next significant action. Thus the operator has time to reconsider subsequent actions programmed for the continuance of the restart. If desired, the operator can reconfigure the programmed actions and, if this does not contravene the overriding logic held by the PMS, it can carry out the modified sequence under the operator's scrutiny.

The start-up of a system can usually be performed in several alternative ways. The safest for a particular status of available equipment is easily identified and should be the primary sequence identified for the PMS for that particular configuration. Any alternative should only be implemented under operator instruction to the PMS, which can advise of any risks or defects in the chosen alternative for the particular set of conditions existing.

The use of the PMS for such activities is ideal, in that it has instantaneous access to all necessary data for correct decision making and it can respond rapidly to any inadvertent occurrences while the restart is in progress. It is essential, of course, that the integrity of the data provided is very high and that the system incorporates a sensitive error detection procedure to ensure that only true data is acted upon. The integrity of the hardware used must also be high, and duplex or triplex equipment may be justified, as discussed later.

Associated systems

Just as a PMS can monitor and control subroutines and safety systems, it can also deal with ancillary systems – e.g.hydraulic, thermal, mechanical – associated with the main electrical system. Thus, for a generating set, the fuel supply system, liquid or gas, water for cooling, waste heat recovery system if provided, and even the steam supply from a boiler system, can all be monitored by the PMS and, if desired, can be coordinated and controlled to give optimum performance. Cooling water systems can have their flow controlled to match loading requirements while maintaining the desired operating temperature on the equipment. This has the advantage of reducing water use at times of reduced loading, and of reducing fluctuations of plant operating temperature which could have adverse effects on some items of equipment. Generating sets provided with waste heat recovery can be controlled to ensure that all items of plant are kept operating within their design parameters and at optimum operating efficiency for the particular loading condition. Where alternative sources are provided, the PMS can ensure that they are available and satisfy the correct supply conditions, and can be programmed, if necessary, to carry out the necessary changeover activities in the event of failure or inadequacy of the existing supply to meet the generating set requirements.

This feature is of particular importance if the PMS is being used to give automatic set start–stop control, as it ensures that all auxiliary supplies are adequate and healthy before initiating the necessary control action. As it monitors the ancillary systems, it can ensure that they are adequately maintained during a start or stop operation.

Equipment loading

The PMS will normally monitor the electrical loading of all major items of plant on the system and make these available on a visual display unit for the information of the operator. It is convenient to show these as a percentage of the unit rated value, as this enables the operator to obtain a rapid assessment of the situation without performing calculations to determine percentage loading. Expressed in this form, these percentage loading values can be used to provide warnings or alarms when preset levels have been reached which might require corrective action. Such actions can be initiated by the PMS if required and, as this action can usually be quicker than that taken by an operator, this may be advantageous from the point of view of system stability.

It is usual to monitor the loading of all generators and main transformers on the system, as this information is essential for maintaining a healthy supply system. However, when the system design is tight and other items of plant are likely to be operated close to their rated capability, these items should also be monitored, displayed and alarmed in the same way. When extending such monitoring it is important to consider the design basis for the system, as it may be unnecessary in some instances. For example, it is usual to match cables to the ratings of transformers to which they are connected, so if transformer loading is monitored and kept within safe limits, the cables will automatically be kept within their rating. This same logic can also be extended down the system voltage levels when a radial configuration is used.

Switchboards, however, are normally selected to have a standard rating in excess of the designed operating condition and this may have a generous margin or may be quite tightly rated. Switchboard loading can change when system configuration is altered, when existing loads are modified or when new loads are added. All these factors can be cumulative, and actual switchboard loadings can increase considerably, often without a deliberate reassessment of the plant capability being undertaken. The PMS, which determines the system configuration from the status provided of all the significant circuit breakers, isolators, etc. can determine the actual loading of each section of busbar, as long as it is provided with the current loading of each connected feeder, either directly or deduced from MVA and measured busbar voltage. Integration of component currents gives a maximum value which is approximately correct. There is an overestimation if the power factors of the component currents are not identical but in most power systems the error is not very significant and is always on the safe side. In cases where the loads are known to have widely divergent power factors it is necessary to provide the PMS with MW and MVAR, which are usually monitored in all generator and main transformer feeders, when the PMS can evaluate an accurate value of maximum busbar current loading in amps. This value is usually presented to the operator as a percentage of the busbar rated capacity under the particular operating condition.

By providing a simple simulation facility within the PMS it is possible to use this function to predict the effects of any prospective changes to the system, whether in configuration, load characteristics or values, or additions or changes to the system. The simulation commences with a copy of the system as currently monitored by the operational PMS and then the operator can use the simulation facility to determine loadings under any other desired set of conditions.

Busbar loading evaluation is normally required only for the main generating or higher-voltage (HV) busbars, as the loading limits imposed by stepdown transformers usually preclude the possibility of overloading lower-voltage (LV) switchboards. However, where downstream switchboards are operated as isolated sections, i.e. with bus sections, this condition does not necessarily apply. PMS monitoring may be justified on such switchboards when the rating has any prospect, either currently or in the future, of reaching the rated values. When split LV operation is provided, it is operationally desirable to transfer section loads to another transformer without major disturbance. This may be required, for example, in the event of failure of one feeder transformer, and is performed by isolating the faulty transformer and closing the appropriate bus section switch to transfer the failed load to the adjacent healthy bus section. The transformer fault detection and clearance time determines for how long the faulted load is reduced to a zero volts supply and hence starting to decelerate all drive motors. Restoration of normal voltage to loads in such conditions can be dangerous, since the decelerating motors generate a voltage which is not synchronous with the supply when restored. Motors can be specified and designed to meet this condition safely, but motors which are not designed for this duty can be damaged in extreme conditions. Such motors can be provided with voltage-sensing protection to indicate when the generated voltage is sufficiently low to permit reclosure, or a safe 'overall' time delay can be provided to ensure that the same safe conditions have been reached. These protective devices prevent reclosure until it is safe.

Autotransfer of transformers, which requires interlocking of transformer circuit breakers on both HV and LV sides with the appropriate bus section switch, is commonly provided using conventional relay logic. When the busbars are split into more than two sections, each provided with its own feeder transformer, this logic can become quite complicated and requires the operator to nominate, in advance, how he wishes the sections and transformers to be reconfigured for each transformer failure. All this logic can be incorporated in the PMS to provide the same safety features, but can also include the system configuration option to determine the mode of transfer without requiring operator preselection. In extreme loading conditions the PMS can also be organized to arrange appropriate switching should some feeder transformers or busbar sections approach their design limits. Such alternatives can affect the fault current levels possible on such switchboards and this factor must be checked.

As a complementary function to busbar, including circuit breaker, current loading, the PMS can also be used to assess prospective fault current levels on the various system components and present this value as a percentage of the equipment rating, providing an alarm if there is a prospect of the rating being exceeded. There are many complicated factors involved in this evaluation, however, which means that a time-variant set of models for each component, adjusted for current loading, would be required for accurate system modelling. This cannot conveniently be incorporated in a normal PMS. However, by using an accurate model of the complete system on a suitable system analyser, it is possible to derive accurate fault level distribution for specific configurations and loadings of plant; these can be used to calibrate a simpler model which can be used by the PMS.

This simple model can be used to assess the effects of minor variations with reasonable accuracy, provided that any additional generators or loads are included in an appropriate manner. Fault level evaluation is only of value on systems which have some items operating close to their limit and where minor modifications to the system could inadvertently cause the rated level of some component to be exceeded. Any major addition or change should first be evaluated by an accurate system study to determine plant suitability and hence derive a new, more accurate model for the PMS and act as a suitable calibration for it.

The PMS estimation must be based on symmetrical conditions

only, as asymmetrical calculations are too complicated. The full system study would perform both, for limiting cases, and ensure that compliance of installed equipment is adequate for all specified conditions. Should there be compatible margins on both calculations, it can be assumed that increases estimated by the PMS for symmetrical conditions and remaining within rating would also be found to be acceptable for non-symmetrical conditions. Care must be taken to ensure that new conditions envisaged do not invalidate this assumption, for example by elimination of significant amounts of cable loss in the system.

Some items of plant have ratings dependent on environmental conditions or ancillary services. If the relationship can be defined and the PMS provided with the necessary additional parameters, it can amend the current rating capability of items of equipment when determining current percentage loading.

Examples of such rating dependencies are gas turbines and their associated electrical generators. The power output of a gas turbine depends upon the temperature of its inlet air; increase of this significantly reduces its output from its design conditions. The output of an electrical generator is based on winding temperature and this can be affected by increase of current loading or reduction of cooling. At rated conditions the machine is assumed to be provided with an adequate supply of cooling medium at a specified maximum temperature, e.g. air at 40°C or less, and under such conditions it can provide its rated current, or MVA at rated power factor. Any increase of cooling medium temperature above the design value will result in a reduction of the current-carrying capability of the generator. The machine can still continue to produce rated MW, however, if it can be operated at a power factor greater than the design value such that the MVA is still within the rated value. Under this condition, other generators on the system would have to combine to make up for the MVAR deficiency of the faulty generator.

Usually, the cooling medium on a generator is air, which is circulated over the windings by means of shaft-driven fans. It is not likely that the flow of air will fail. However, most large generators do not use free ambient air from the generator room for cooling, and employ some form of heat exchanger in the air circuit. This enables the generator losses appearing as waste heat to be removed conveniently from the generator room. Deterioration of heat transfer in such heat exchangers, due to fouling, leakage, reduced secondary coolant flow or temperature of secondary coolant exceeding its design value, will result in increase in the temperature of the primary coolant responsible for removing heat from the windings and this in turn can result in reduced current-carrying capacity.

By monitoring the relevant parameters, including the coolant systems, the PMS can ensure that the system is functioning correctly, or if not it can deduce the theoretical design capability of the generator sets, both prime mover and generator, for the particular conditions and use this value when presenting percentage loading values to the operator.

There are conditions which can arise in service where the generating plant is unable to meet its theoretical rating because of deterioration such as fouling of turbine blades or cooler tubes; this would result in an optimistic value of capability being derived by the PMS by using the above procedures.

It is usual to provide equipment-monitoring devices to warn when plant is reaching its safe loading, usually based on operating temperatures of the significant component. Such protection is independent of actual rating and is intended solely to warn the operator and take corrective action to save the plant from damage. This equipment can be used in conjunction with the PMS logic for condition monitoring, referred to later, and can determine the amount by which the plant rating has been effectively reduced due to fouling etc. This value will give a more accurate assessment of the generator set capability than the theoretical ideal value.

The procedures described for generating sets can also be applied in appropriately modified form to all other items of generation and distribution plant, such as transformer, switchgear etc., to enable suitably accurate values of rating capability to be obtained. It is not usually necessary to provide such accuracy for every system provided with a PMS, but the practical limitations of assuming ideal conditions should have been borne in mind when deciding whether to regard plant rating as the nominal value (assuming all qualifying requirements are being met), the theoretical rating (adjusted for the plant operating ideally under specified parameters which differ from those on which the nominal rating is based) or the current practical capability (based on the safety limits specified by the manufacturer and designer of the plant).

Another example of using the PMS is in the event of failure of a generator AVR, which normally results in the protection system tripping the set off the busbars; this can have a serious effect on an island system. As the set remains capable of providing rated MW output, it is customary to provide a simple standby excitation system which is manually controlled. This enables the set to be put back in service under operator control.

By incorporating an 'excitation follower' system, which automatically adjusts the value of the standby excitation to match that being supplied by the AVR system, it is possible in the event of an AVR failure to transfer automatically to the standby system, leaving the generator operating at the same value of MW and MVAR, and also to provide an alarm to the operator that such failure has occurred.

Until the AVR has been repaired and reinstated, the generator will remain under constant excitation control and will provide an

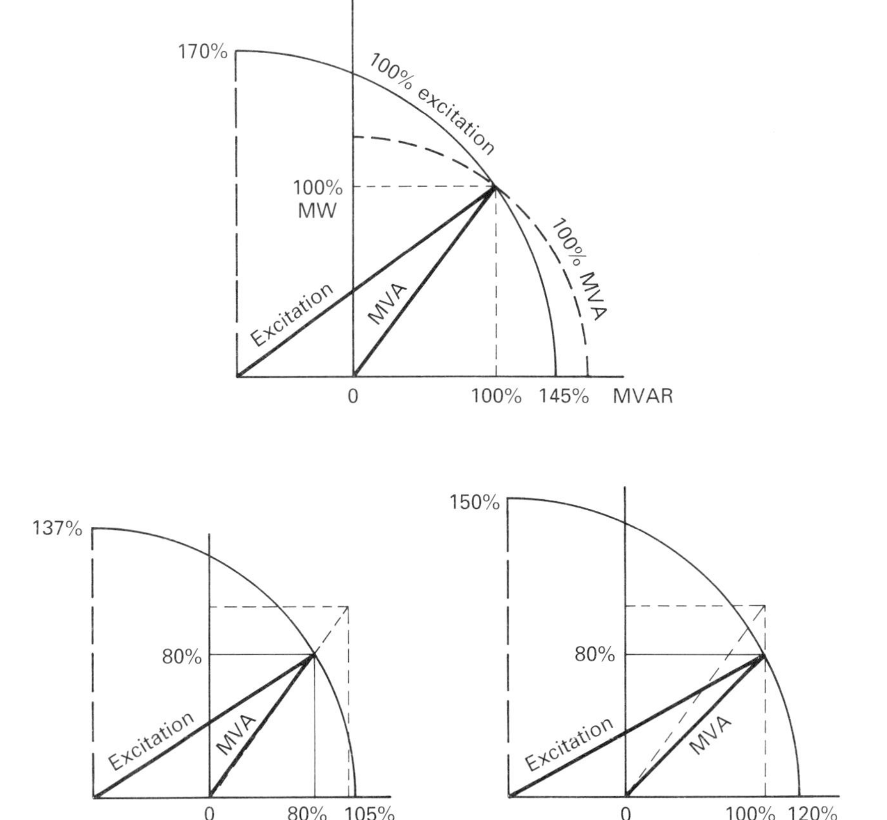

Figure 4.1 *Synchronous generator MW–MVAR–excitation relationships*

amount of MVAR related to the MW being demanded by the PMS system via the set governor by the generator excitation characteristics. Figure 4.1 shows a typical generator operating at rated design conditions, and with this value of 100 per cent excitation it would provide about 145 per cent of designated MVAR at zero power output, and would allow the generator to produce up to 170 per cent rated MW before losing synchronism. In the event that the set reverted to fixed excitation control when operating at 80 per cent rated MW and at constant PF, it would only provide 105 per cent MVAR at no load, and have a pull-out value of 137 per cent MW. Had the excitation control been set for constant MVAR, the corresponding values would have been 120 per cent MVAR and 150 per cent MW.

Both conditions leave the set inferior to the normal conditions and could require other sets on the system to be overexcited to maintain normal operation over the whole range of possible system loadings. As an alternative it would be preferable for the operator to adjust the standby excitation to 100 per cent following AVR failure and leave the set capable of normal full-power operation. Under all reduced-load conditions the generator would be providing more than its share of MVAR, thus relieving other sets on the system. This condition could cause difficulties in the event of the MW load demand falling to a low value but in such an event the PMS system would warn the operator to shut down one or more sets, leaving the remaining sets operating at economic loadings; obviously the set with the faulty AVR would get preference for shutdown.

Protection

Conventional protection equipment operates to provide a positive, usually single, action to compensate for some detected condition which could result in damage to equipment, dangerous maloperation of plant or installation or hazard to personnel. Such devices are essential to maintain adequate safety levels in a system or plant and to ensure that installed equipment is kept within its designed operating limits.

However, when considering a total system such as an island generating system, any abnormal condition that occurs may have significant effects on different parts of the system and items of plant; while each of these could be protected by selected protection equipment, a predictable series of events could develop from any single occurrence and rapidly lead to widespread damage or shutdown. This is a particular situation in which a PMS can provide a significant protection function by initiating such system control functions as could reduce the extent of the disturbance and consequent interference with continued system or plant operation.

The advantage possessed by a PMS is that it has available a wide range of system parameters and from these it can be determined what logical action is best able to compensate for some measured abnormal condition. A protection relay usually measures only one, or sometimes two, parameters and it must take a specific action on any excursion of these outside the acceptable working range. Usually the action taken is drastic, since it must be adequate to deal with the worst condition that could cause the measured abnormality, and if this is in fact the result of some less serious condition, some equipment or plant may be shut down unnecessarily.

One problem encountered in island systems is that caused by generator fault contribution to the system. This depends on the number of generators connected at any particular time and can often vary significantly. During start-up conditions only one generator will be connected and the possible fault current will be low, but since the protection system must protect against the greatest corrected value it is often difficult and sometimes impossible to select a setting which will give adequate detection for all possible operating conditions.

A similar problem can arise when a system can be split and used as one or more subsystems, but since the PMS will have data on the number of connected generators and on the system configuration it can determine what relay setting would be appropriate for such conditions and can be arranged to use this information by modifying the relay setting directly, or selecting a trip signal from alternative relays with the appropriate settings.

A logical extension to this approach is to include the relay function within the PMS software by providing it with a suitable current–time set of characteristics. Since the system currents are already available to the PMS, it can select the appropriate relay characteristic and settings and then trip the appropriate circuit breaker if the set current level is exceeded.

Many protection relays now employ their own microprocessor to give a greater range of flexibility: whether they are interfaced with the PMS, or to the equivalent software included in the PMS, is a commercial choice which is readily available.

Another aspect of PMS use for protection is where several control functions may be advantageous as well as that of tripping a circuit breaker to isolate some fault. Thus when a major fault occurs on an island system the frequency will tend to rise as a result of disappearance of electrical load with voltage, and the prime movers will tend to reduce fuel to compensate. However, as soon as the fault is cleared the load will reappear, and since any induction motors will have started to decelerate, the restored load will differ from the prefault value and may have a much higher MVAR demand. As the motors will have to reaccelerate, there may also be a significantly increased MW demand. These will result in a reduction in generator voltage and deceleration of the prime movers, reducing the overfrequency to normal and then to underfrequency.

The PMS, if informed that a specific type of fault has occurred, can determine the optimum action to be taken. Thus, by inhibiting the fuel reduction to the prime movers, it will enable them to accept a greater load after fault clearance without having to wait for a fuel increase which is usually limited to a ramp rate determined by other conditions. The PMS can initiate optional actions based on existing generation capacity and system loading and plant lost for each fault condition. Suitable earthing of the system or subsystems can also be arranged by the PMS. Details of extending the normal protection system to include associated functions in this way are discussed in Appendix 2; to obtain full benefit from this, system analysis studies must be carried out to determine behaviour during both transient and fault conditions.

Condition monitoring

The PMS can be used to present the various operating parameters of the system and components of plant relevant to its condition and can give their relationship to the normal design values, as described in the section on equipment loading. However, to function as a condition monitoring system the PMS must determine whether the items of plant are performing correctly at the appropriate loading. Thus by measuring the temperature of transformer windings, insulant, etc., together with the cooling medium and ambient air temperature, and taking into account whether forced-fan cooling is operating, or, with liquid cooling media, such as water, by measuring flow and temperature rise, the PMS can check whether the performance corresponds to the actual loading, using factory or on-site load–temperature tests as the reference criteria.

Prime movers can be monitored in a similar manner provided that adequate accurate data are available together with the proven design operating conditions. Thus measurement of the inlet air temperature and the exhaust gas temperature of a gas turbine enables these values to be compared with the design operating values for the measured generator electrical output. Discrepancy between the values can indicate the deterioration of machine performance. Such simple parameter checking assumes that all other parameters are constant, which is not always the case. However, if these other items can also be measured and checked with the norm, compensation can be made for them; for example, fuel quality can have significant effects on machine performance.

Other parameters, such as lubrication system flows, pressures and temperatures, can give indication of incipient machine problems and are frequently used for condition monitoring. Likewise, rotor vibration levels can give evidence of possible problems, particularly if data are available on velocity as well as magnitude in the three planes relevant to rotating shafts.

There exist sophisticated devices which can carry out selected condition monitoring functions. These are frequently processor-based and their output data can be passed to a PMS for integration with other relevant information to give the operator maximum information regarding the performance of the plant. This arrangement can be very convenient, as in some cases, instead of having to program the PMS with a range of operating parameters together with the analytical software, it is possible to use what has already been developed and proved reliable in service in other installations.

Most prime movers have some degree of condition monitoring built into their control systems but they can also be provided with quite elaborate equipment for this function. The most economic overall solution which gives the necessary degree of accuracy should be assessed before including it all in the PMS.

Maintenance programming

A feature closely related to condition monitoring is that of maintenance programming. A PMS can be used to monitor and organize this if required. Maintenance should include both preventive and corrective activities. It is also useful to include a development function which can be used to suggest changes to plant design features or modes of operation which can reduce outages in the future.

Preventive maintenance is based on condition monitoring as far as equipment deterioration is concerned. However, the manner in which plant is used in service can cause unnecessary deterioration and risk of failure, and a PMS can be used to monitor such conditions and either warn the operator, or record the operation and subsequently analyse the records to highlight a factor which could be improved.

Data available to the PMS can enable an accurate reconstruction to be made of conditions prior to and immediately preceding plant failure and can also indicate what operations had been carried out relevant to the failure. Analysis of this, together with the actions taken by the various protective systems and shutdown procedures, can determine whether the failure could have been avoided or detected prior to severe damage being caused. Such an examination can lead to the most suitable corrective action being taken on future similar occurrences. It can also provide information leading to replacement of components likely to fail in the future or even to the redesign of subunits or parts to avoid similar trouble in the future.

The PMS monitoring and recording facility can prove very useful for such functions, as it can store all data relevant to maintenance, correlate it and present it when requested in any desired format at either local or remote locations.

Satellite systems

Island system plus satellite

With this arrangement all the features described above are relevant but an additional factor is introduced by the impedance of the interconnector between the generation location and the satellite load. This results in a voltage variation at the satellite location, depending on the interconnector impedance and the load carried, which can have serious effects on the load behaviour if not controlled. While it is desirable to keep the receiving-end voltage within the same working range as the sending-end voltage, this is not always practicable; however, it is still necessary to keep the load voltage within the range necessary to ensure satisfactory operation of the components of the load. Different types of load require different voltage ranges, and it is necessary to meet the requirements of that having the smallest range. This has a significant effect on the type of voltage control provided and can prove a severe limit if step control has to be used, such as a transformer tap-changer. The receiving-end voltage regulation is most seriously affected by the MVAR loading on the interconnector, since the line impedance is usually predominantly reactive. However, special conditions can arise where very long lines have to be used. Accurate values should always be determined before deciding on a suitable control system. Figure 4.2 indicates the relative effects of the MW and MVAR loads, independently as well as combined, on the sending-end voltage required to ensure constant rated receiving voltage.

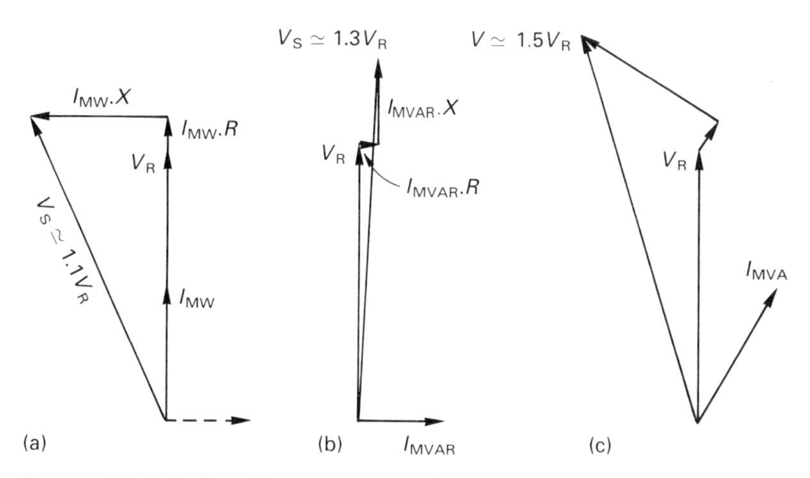

Figure 4.2 *Relative effects on MW and MVAR loads on voltage regulation*

When the line voltage regulation is small it is possible to compensate for this by using a line transformer with an on-load tap-changer. This should be installed at the generating end of the line so that the transformer primary can be maintained at busbar voltage, resulting in optimum performance of the transformer and maintaining the line voltage at, or above, the nominal receiving-end voltage.

With long or high-impedance lines it may be impractical to operate the line at the nominal receiving-end voltage: it will be necessary to install a transformer at each end of the line and operate this at a voltage higher than nominal system voltage, consequently reducing the effective line impedance. It is then customary to use the tap-changer on the sending end to adjust the high voltage on the line to above the nominal line voltage, so that the line voltage at the receiving end is slightly below the nominal value. The receiving end transformer tap-changer can then be used to adjust the receiving-end load voltage to be within the required operating range.

The PMS can control the tap-changers as required to give the desired operating conditions as the load changes. It can also include a pre-emptive function, if required, to minimize the transient voltage drop resulting from switching a large load at the receiving end. When the request to start the load is made via the PMS, it can raise the line and receiving-end load voltages by an amount determined by the load to be switched, and the consequent transient voltage drop will not result in such a severe value of voltage reduction at the load. When the transient effect has disappeared, the PMS can then regulate the steady-state voltage to the desired value.

All the other PMS functions can be applied in this operating mode but it is necessary to modify the simple load-shed logic that can be applied to a simple island system. This is necessary because the effect of shedding a satellite load is affected by the impedance of the interconnection link, including any transformers, which depends upon the value of tap-changers in current use, as well as the actual value of load being transmitted. The data required to calculate this are available to the PMS, which can make the adjustments to the nominal load being shed to determine the actual effect on the generation system of the particular load considered under the system conditions currently prevailing.

Impedance-connected island systems

When two island systems are interconnected through an impedance, the combined system at the instant of synchronizing will be steady-state stable and operate in the same manner as the two independent systems. However, as soon as any system parameter or condition is changed, the dynamic behaviour of the two systems becomes interactive. If they are not compatible, the systems can become unstable and in the extreme case this may result in the systems being separated by the protection system provided.

The use of a PMS can improve conditions, but as it is usually necessary to operate the two systems independently during start-up, or in an emergency, it is customary to provide a PMS system for each island system and to utilize one or other of these as a master PMS for the combined system when the islands are interconnected. The second PMS can be used as a standby unit if desired.

By using a single PMS for both island systems when interconnected, it is much simpler to improve stability, as similar forms of control can be used for both generation set types, and the separate load-shed facilities can be integrated into a single system. Thus steady-state control can be carried out rapidly and accurately. However, transient stability still has to be checked to ensure that any dynamic mismatch between the two independent systems is not so great that the disturbances that may be encountered will result in instability. It is not practicable or necessary to get a precise match between all dynamic responses, but it is desirable to get them to be reasonably similar. This aspect of operation requires a careful transient system study to determine optimum system parameters.

In this operating mode the PMS can utilize all the features described for a single island system and can combine all the generating sets in any nominated order of preference and with any desired form of loading on individual sets – base MW load, proportional MW, constant PF, constant MVAR or proportional MVAR – depending upon the type, condition and performance characteristics of each set. Similarly, all the sheddable loads can be included in a single load shed sequence of priorities.

Two island systems are often interconnected to give improved standby capacity in an emergency on either, and each will have enough generation capacity for its own load. In this way, disconnection of the two will leave the two separate systems stable and able to continue operating normally.

However, it is more usual to use other criteria to determine the operating mode normally used. Where different types of generating sets are used, some will benefit from running at a fixed load both in terms of fuel economy and reduced servicing requirements, and these will be high in the priority rating of running sets. Less efficient sets can be used to provide for the power fluctuations of the load and also to provide redundancy if required. The advantages gained by such operation have to be offset against the extra cost of the interconnection line, which may have to carry a greater load than if it were provided only for transmitting a limited amount of emergency load, and also against the reduced transient stability of the system with the predominance of generation located at one load centre. These disadvantages can be partially compensated for by providing a suitable load-shedding system to assist transient recovery following a major system disturbance, in conjunction with a fast protection system which will detect such a fault, take action to isolate the fault rapidly and also activate the PMS to implement the necessary control actions to allow the system to recover most effectively.

Another aspect of operational control which may be relevant is the optimization of load transmitted between islands to reduce losses, and hence costs, and to ensure that plant operates well within its design capacity. For example, generators operating at reduced MW loading can be used to provide MVAR locally at little cost and avoid importing the equivalent MVAR via the interconnector, thus reducing transmission losses and easing the voltage regulation problem on the line, giving a double saving of MVAR required from the sending-end generators.

Another feature that can prove useful in special cases where tap-changer steps available are coarser than desired is to adjust the two island generation voltages to values slightly above and below nominal respectively, but still within acceptable plant operation limits. The accumulated difference can be finely adjusted, via the PMS if required, and thus give closer regulation on line voltage control.

When impedance-connected island systems have satellite loads, the control logic follows the principles described for each single island system together with that for the interconnection control, and presents no difficulties for the PMS other than those described. However, should there be two separate impedance interconnectors between two island systems, forming a closed ring, a completely new problem is presented if total system control is required by the PMS. This can be further complicated if satellite loads are fed from the ring rather than directly from one of the island generation busbars.

Ring-connected island systems

The significant difference of such a system, as opposed to a simple radial system, is that the magnitude of MW flow in the two alternative circuits is determined by the effective impedances of these circuits and is affected by the magnitude and location of the various MW loads and also by the magnitudes and sources of the available generator MW outputs. However, the MVAR flow is affected by the voltage distribution in the system, including the generator busbar voltages, system impedance voltage regulation and any transformer ratios, including tap-changer control. There is a measure of interaction between the MW and MVAR flows, since both affect the line impedance voltage drops, and also the actual load voltage can affect the MW and MVAR demand at that location.

Before a ring system is closed it behaves as an impedance-connected multiple island system with satellite loads. Closing the ring at any location merely means adjusting the two open-circuit voltages at that point, since the whole system is synchronous. Closing the ring, however, is significantly different from closing an interconnector between the two island systems, as MW will immediately start to circulate as determined by the current system impedance and load distribution. This flow will cause immediate line voltage changes, resulting in MVAR flows in the ring which will normally reach a stable steady-state distribution of MW and MVAR. At this point the PMS, which will be advised of the component values of parameters, can start to adjust these according to its nominated priorities and leave them within its nominated range of values. Subsequent changes of load, generation or system configuration will be dealt with in a similar manner. Any fault condition will, however, have quite serious consequences and it is essential that the PMS is notified as quickly as possible of the fault: being aware of the actions being initiated by the protection system, it can initiate its own control actions which are intended to minimize the disturbance resulting from the fault and restore as much as possible of the residual system to a stable operating condition.

The logic required to program the PMS for the above is inherently simple in principle but complicated by the number of possible conditions which must be considered and analysed to confirm the most effective actions required. It is also necessary to establish an operational priority for the various loads involved, and to determine what limits are considered permissible for the various system parameters. Obviously, transient system stability is one of the most important criteria involved here, and the operation of the system in a healthy mode can affect its ability to recover, if correctly controlled. It is important, therefore, to establish the best operating conditions to provide system stability of the highest order. The PMS can be advised of these and either control the system accordingly or alternatively advise the operator of such information.

When the ring system is operating abnormally – i.e., with some major item of plant out of service – the desirable operating modes may be significantly different and the PMS can also be advised of these and take the necessary action.

Satellite island system

When an island system is interconnected with another, much larger system the operating constraints are altered significantly and the mode of control is quite different from that of an isolated island system (Figure 4.3). The large system, operating at constant frequency, can act as a source or sink of MW to or from the island system and the value is determined solely by the local generator governor control which determines the total island MW generated. The difference between this and the total island MW load then has to be compensated on the interconnector to the main system by importing or exporting MW. As the frequency remains constant, the local generator will produce a constant value of MW, and any change in local load will only affect the imported or exported MW value, unless special governor control is used to regulate this. The mode of control should be determined by the economic values of imported and exported power and the manner in which the local MW load varies.

The MVAR circulation in the interconnector is more complicated than the MW, which is virtually determined by the generator set governors. In its simplest form, if the island system were connected to a constant-voltage busbar on the main system through a fixed-value impedance then, if it was operated at constant local busbar voltage, the interconnector would be constrained to transmit only a limited range of MW and MVAR (see page 41). Figure 4.4 shows possible MW and MVAR flow conditions.

In practice, it is usual to provide voltage regulation devices on the main system interconnector circuit which can, in effect, change the impedance voltage drop, e.g. by adjusting a transformer tap-changer to compensate for either load throughput or actual

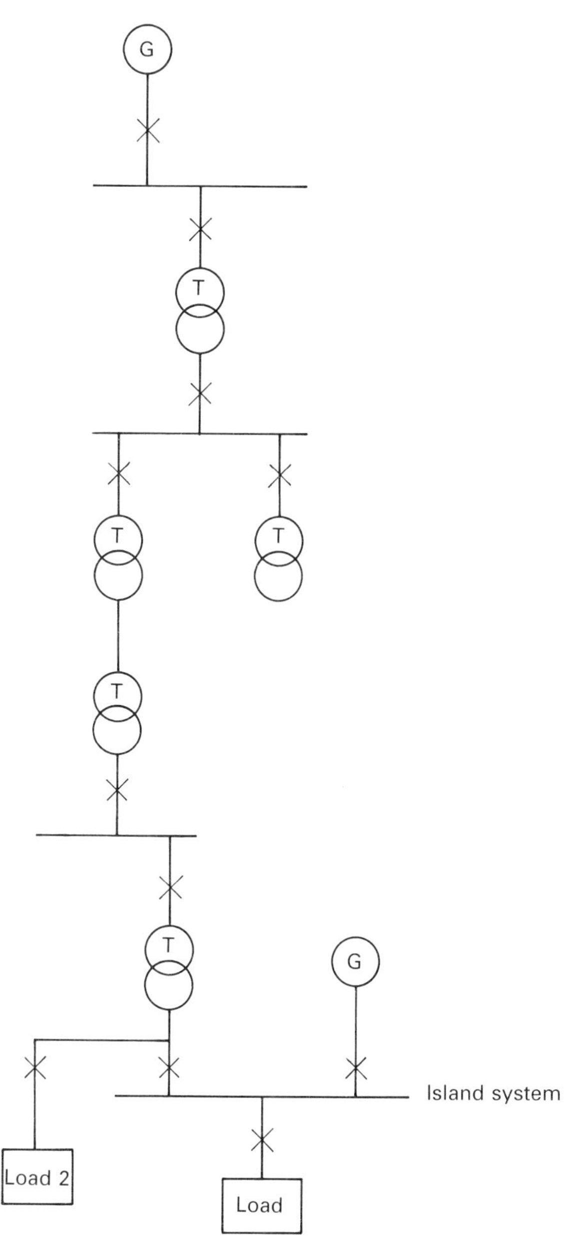

Figure 4.3 *Typical satellite island system interconnected to large system*

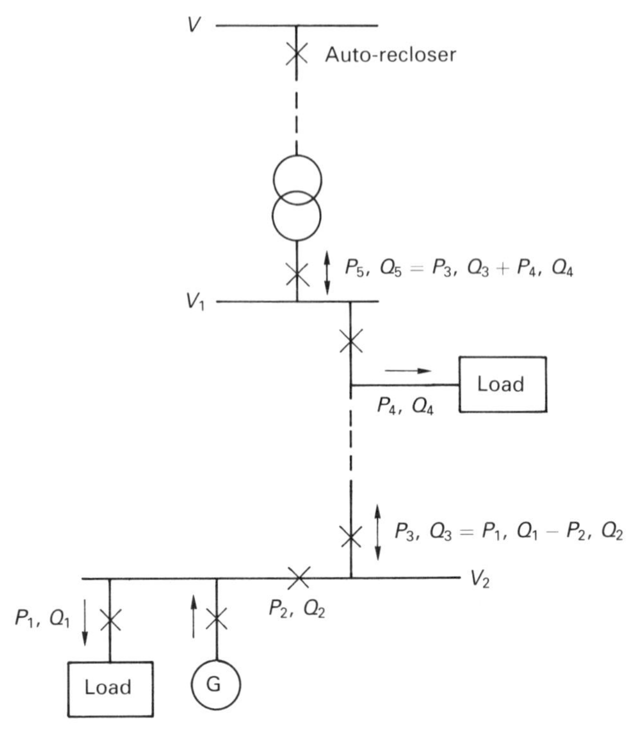

Figure 4.4 *Possible MW and MVAR load flows for satelite island system*

secondary voltage. However, the effect of such devices on the island system generation is not intended to optimize this, but to regulate the main distribution system within prescribed limits, and the function of the island generator AVR in adjusting excitation to produce constant busbar voltage results in the MVAR generation being determined by parameters outside the island system. This can result in the island generators being overloaded irrespective of the demand of the island loads.

Under such conditions it is usual to utilize the AVR to control the generator excitation to regulate the MVAR in a mode likely to keep the generators within rating and the local busbar voltage within normal operating tolerances. The best method of doing this is dependent on the nature of the local load and its variations, and can take the form of regulating the generator PF to a nominated value, or the generator MVAR to a constant nominated value. Alternatively, control can be applied to keep the imported/exported MVAR to a constant nominated value.

In satellite island mode of operation it is necessary to determine the desired operating criteria quite carefully and accurately to ensure the plant is utilized effectively and economically within its design limits. To do this it is also necessary to consider the implications of voltage-regulating devices for the interconnector together with its MW and MVAR transmission capabilities.

It is obviously convenient to use a PMS system that can be programmed according to the desired criteria and which, when provided with the necessary system parameters, can perform the necessary system control functions for MW and MVAR. It can also provide all the additional features previously described as necessary for the satisfactory operation of the island system. In the event of the island being isolated from the main system by disconnection of the interconnecting network, the PMS can automatically restore all the conditions necessary for normal island system operation, including any system reconfiguration, necessary earthing and protection adjustments, and optimum MW and MVAR control and distribution.

Integrity of PMS

Possible functions

When a PMS system is adopted it is necessary to identify clearly its status and scope. It may be used to provide several alternative functions or combinations of functions, as set out below, and the way these are used will affect the level of integrity required in the PMS.

1 The PMS may be used only to provide information to an operator who functions in the conventional 'manual' mode; that is, the operator initiates the necessary automatic systems provided for the starting/stopping and control of generator sets, and open and close circuit breakers to configure the system to meet the existing or prospective load requirements.
2 In addition to the basic data normally provided by instrumentation, other parameters of interest to the operator can be provided continuously, at specified time intervals or on specific request by the operator.
3 It is a simple matter to record any desired data on either a short-term storage basis or as a permanent print-out. Selected data can

be recorded at preset time intervals as required. A PMS is particularly useful for recording alarm signals from the system or plant, as it can give indication and warning at selected locations where urgent action should be taken. Its recording procedure can identify all alarms occurring during a complex disturbance and this can be important when subsequently trying to analyse the sequence of events to evaluate the real causes of failure or maloperation.

4 The PMS can operate on any monitored data to carry out simple computations such as averaging loads, totalling values or integrating functions (e.g. MW hours) and so relieve the operator of time-consuming and tedious tasks when compiling system operating reports. It can also be used to perform large or more complex computations such as evaluating system loading capacity which would require an operator to use some computer facility if carrying out the same task himself.
5 Since the PMS has access to all relevant data, it can be used in a supervisory capacity to compare values with nominated criteria and provide alarms or warnings to the operator when satisfactory operating margins are no longer being met.
6 Most PMS units can interface with standard telecommunication systems and can be arranged to transmit any or all of the data handled to any remote destination. This can be in the same form as it is presented to the operator locally, or in the form of printed out data, as required.
7 Control functions can be performed by the PMS on specific instructions from the operator using a standard form of keyboard. This arrangement can result in a reduction of space requirement for conventional operator's desks, panels, instrumentation, etc. At the same time, duplicate PMS keyboard/VDU operator units can easily be installed to permit control from alternative locations. The integrity of the control functions would be the same as when normal operator initiation was used; that is, it would be included in the control system itself used for the particular function, such as set starting and generator synchronizing.
8 When the PMS has access to the necessary data defining the system parameters as well as to system control function, such as set starting, it only requires the addition of the necessary logic to organize it to perform the function that is usually initiated by the operator manually. The operator can still override the actual

sequence, if desired, make the necessary assessment from the PMS data, and initiate the control action via the PMS. This feature is valuable when commissioning the system, or investigating undesirable operating modes, should they be encountered.

9 As systems become more complex, the data analysis may indicate several necessary control actions, of which the different possible sequences may have significantly different effects on the system during transition conditions. To leave the operator to analyse the alternatives and to initiate the correct sequence of actions at appropriate times may be unreasonable or, in some instances, quite impractical. This is particularly so when taking the necessary rectifying control actions following plant failure, or system fault, which must be done accurately and rapidly if system stability is to be maintained. Provided that the necessary system analyses have been carried out to identify critical conditions which can arise in such circumstances, and the interaction of the various control actions has been correctly determined, the PMS can be used to perform the necessary actions following any conditions indicated to it. This requires full integration of the protection system with the PMS. The PMS must differentiate between normal, healthy disturbances and fault conditions, as the same system parameter criteria changes will require quite different control responses for these alternatives.

10 These additional capabilities of the PMS can be used for training purposes by locking out the various control actions and leaving the operator to initiate manually the action specified by the PMS, which he should have determined for himself from the data available. This is not practical for fault conditions, and for such a facility the next capability can be provided.

11 A simulation facility can be provided with a PMS system which can contain all the logic required for on-line operation and be provided with the live input data. In the simulation mode the resultant actions can be determined and, if desired, the effects of changing input data can be determined.

When the interaction of the system parameters due to some control changes results in the need for subsequent control actions by the PMS, it is necessary to incorporate a realistic 'system model' in the simulation mode to enable a reasonable assessment to be made of the PMS behaviour under complicated system conditions. Obviously the accuracy of the prediction depends upon the accuracy of the system model used and it is customary to validate this against the results of the more accurate transient stability studies performed for the system.

When only steady-state conditions are being simulated, a practical accurate model can usually be derived using normal assumptions. However, it is normally impracticable to attempt to include accurate dynamic or transient models in such a simulation because of the complex interactions between all such components. Such behaviour can only be accurately determined from a comprehensive transient system study.

12 Data presentation by the PMS is an important function but must be well organized to avoid the operator being saturated with so much material that the important items become difficult to find. It is just as important to omit irrelevant data on which no action can be taken as to provide all the data essential for making a specific decision or group of decisions. The amount of data that can be presented on a VDU screen is limited, and often attempts to include too much reduce clarity and simplicity. Care must be taken that sufficient information is provided to make each separate page useful on its own without being overfilled with data, which makes it confusing or difficult to assess quickly.

Each page of presentation should be prepared for a particular function exclusively, with reference to other associated pages for additional information only where necessary. Thus the complete system configuration is usually impractical, but the generation busbars and associated voltage equipment can be presented on one page and LV sections of the system presented on separate pages, which only interface at a few points on which full data can be provided on corresponding pages to ensure continuity.

Other functions, such as load shedding, can usually utilize a single page to include all drives available to the system, their status, such as running or stopped, load, location in the shed sequence, autoshed, reset, etc. It is also convenient to incorporate reacceleration data, if relevant, as the overall status of all associated drives can be presented simultaneously. Other special pages can be dedicated to equipment such as status and performance of generator sets and can include alarms and warnings of limiting parameters being reached.

13 There are many other functions for which the PMS can be used, as described elsewhere, but in general they do not introduce any factors relevant to integrity other than those referred to above, insofar as the PMS itself is concerned. However, the integrity of associated auxiliary systems or equipment should be evaluated in the same way as for the PMS and should be of the same order of integrity.

Electronic control equipment

Microprocessors have been used extensively for system surveillance and have been developed as very high integrity devices. They have proved suitable for such critical functions as fire and gas detection and shutdown systems by reliable service over several years. Recognized standards of performance exist for both the necessary equipment and the manner in which it should be installed, operated, supervised and serviced. Many forms of equipment and system are currently available, each with its particular advantages for different applications.

Various levels of integrity can be provided, depending upon the particular requirements of each situation: the greater the level of integrity required, the greater will be the complication of the equipment and consequently the greater the cost. This cost must be offset against the financial consequences of maloperation of the PMS, which depend upon the functions it performs and the economic consequences of disturbance of the power system that is controlled.

Thus, if the PMS is being used as a back-up supervisory or alarm system, its failure or maloperation need not be serious as long as the alternative system remains viable. However, if is being used as the principal alarm system, then the required level of integrity is determined by the function of the power system; for example, if it is being used for the propulsion and steering on a ship, a very high level of integrity is required.

The following distinct areas are involved in determining the level of integrity of any PMS control system:

1 The hardware and its power supplies.
2 The data provided for the PMS.
3 The program (software) used to apply the necessary logic to the data.
4 The interface between the PMS and the system controls.

It is also necessary to examine all the components of the power system, including all generation and distribution components. The following procedures can be applied, in principle, to all aspects of the complete system, including all its control elements.

Hardware

The reliability of all hardware depends upon the capability of each component to perform all its necessary functions under all practical operating conditions, normal and abnormal, and also its suitability for all environmental conditions occurring during transit, storage and erection, as well as during operation, as discussed previously. It must also be capable of full quality control throughout manufacture, including component selection, inspection and testing, and be fully accessible for inspection and maintenance when installed.

All devices incorporated into each component must be quality controlled to ensure compatibility with regard to reliability and operation well within design limitations under all conditions of operation. This will result in acceptable MTBF (mean time between failures) values being obtained for the equipment.

The power supply for all hardware should preferably be from a UPS(uninterruptable power supply) to permit continued operation in the event of mains power failure, but it is often more convenient to provide two alternative power sources with an automatic changeover from normal running direct on the mains to a standby UPS, to cover the time required to restore the mains following a failure. This reduces the continual load on the UPS. It is essential that the power supply is 'clean' and does not produce excessive voltage disturbances which could damage the control system; otherwise a UPS must be used or a suitable filter must be added to the mains.

The enclosure of each component of hardware must provide the necessary level of protection for all contained components to keep them in a suitable environment – i.e. clean and adequately ventilated – and also isolated from excessive shock and vibration effects. It must also be convenient for handling during erection, and have facilities to make inspection and maintenance simple.

When a particularly high level of reliability is required, this can be obtained by providing two complete duplicate processor units, each with its own power supplies, and having an automatic changeover, should the master unit fail, to the standby slave unit. Both units should be kept normally on-line, and capable of functioning when required.

Care must be taken to prevent interaction between the two units to minimize the risk of any fault in one affecting the other, and it is customary to install them in physically separated and protected, e.g. fireproofed, compartments.

Such duplex systems are in common use for high integrity and have proved their reliability in service. When such equipments utilize common input signals or output functions, this does increase the risk of common mode failure. Provision of separate, duplicate signals to each system can avoid this risk by eliminating interaction between the systems; this solution does increase integrity, and should be adopted at least for those signals which are essential for the safe operation of the power system.

A further improvement can be obtained by using three processors with independent signal sources and adopting a two-out-of-three voting procedure to determine which, if any, single signal is faulty. The actual level of improvement obtained with this system is doubtful, as the increased complication of circuitry itself introduces a level of risk that offsets the advantage gained.

Data

It is essential that the PMS receives and transmits accurate signals on healthy circuits. Circuit continuity can be checked by conventional methods: input signals depend on the accuracy and manner of application of the various sensing devices; output signals rely on correct software logic correctly applied.

It is usually convenient to use existing sensing devices where these are available and of sufficient accuracy, e.g. electrical circuit voltage and current instrumentation transformers. It is customary to use the same devices provided for visual instruments, as this gives a quick means of confirming that circuits are healthy during commissioning and simplifies calibration of the PMS value with the meter reading. Special analog signals must be obtained from devices of the necessary accuracy, which must be installed for use by the PMS if they do not already exist.

All system signals have a practical range of operation and this is normally used to detect any abnormal condition indicating some form of failure. However, during transient conditions some functions exceed the normal, healthy range of values and allowance must be made so that such excursions are not interpreted as a signal failure. It is important that the behaviour of sensing devices can be accurately predicted over this abnormal range of operation and that due allowance for this is made in the software logic.

Status digital signals, such as circuit breaker open/closed/out of service, utilize conventional auxiliary contacts as used for normal safety and control. For the more important functions, such as safety shutdown systems, the most reliable should be used – i.e. those controls mounted on the circuit breaker itself and not contacts provided on follower relays.

Many signals are opposites or alternatives; for example, switches usually have contacts which indicate open and closed states. The PMS, when provided with both, can check that both signals change together, indicating a healthy state. This condition also arises when import/export functions such as MW and MVAR exist. Should both signals appear simultaneously, this indicates that one is unhealthy. Usually the PMS will also receive a current signal for such a circuit and, from this, actual confirmation can be obtained of the correct switch state, since a current signal will only be obtained when the switch is closed. Thus the faulty signal can be identified.

Such comparative logic can be extended to other important functions on the system. Thus integration of the MW and MVAR on all circuits on a common busbar should indicate zero, and this logic can be applied to complete satellites and island systems. Where considered necessary, as on generator circuits, the values of MW and MVAR can be combined to determine generated MVA, and from the values of generator current I and voltage V this MVA can be confirmed, indicating that all four signals are healthy. If parity is not indicated, the value of voltage is easily confirmed, as it determines all system values and the current should be within the generator capability range. The MW value also has to be within the set capability and should be the basis for PMS control of set load sharing.

Should the three values indicate normal operation, the conclusion would be that the MVAR signal was faulty.

This form of cross-checking values gives a ready means of determining healthy circuit values and should be applied to all parameters that are used for primary control functions on the system, such as generator MW loading.

When an apparent choice of parameters is available for an input signal, care must be taken to use the one giving positive indication of the actual function required and not necessarily the most convenient one. Thus the indication, on a generator set, of prime mover full speed does not prove that the generator is providing rated frequency: it may be unexcited or have a winding fault, or may even not be rotating, because it has been physically disconnected. Conversely, although the frequency generated by a generator operating on open circuit gives a value proportional to the prime mover speed, a generator connected to a system operating at rated frequency need not be connected to a prime mover running at rated speed, as the generator may be operating asynchronously.

It is frequently necessary to supplement important signals with other relevant signals in order to identify the correct significance of the actual function, and hence the condition of the equipment or system, to determine the correct control action indicated by the signal. It is not acceptable to assume that all relevant conditions are normal and healthy when deducing the implication of any variable: for example, a sudden increase in system frequency may be due to a sudden load reduction, in which case it would be associated with a rise in system voltage. However, it could be the consequence of a system fault, which would be associated with a voltage reduction. Each condition requires a very different control action by the PMS.

In addition to ensuring that all the necessary accurate signals are available to the PMS, it is essential that these are carefully calibrated during commissioning and also again at regular intervals to check that no deterioration has occurred in the sensing devices used.

Logic

Preparation of satisfactory software logic requires a thorough understanding of all the interactive functions involved in the system, including both electromagnetic and electromechanical phenomena, together with an accurate understanding of the behaviour of the electrical loads under all possible operating conditions, normal and abnormal. It is impracticable to include an accurate model which can represent the response of all system component functions simultaneously; in fact, simultaneous consideration of the behaviour of different functions could mask effects which the PMS should react to. Determining reaction to combined actions requires assumed conditions under which the independent functions have changed, which excludes the possibility that any particular function is varying as the result of a separate cause. It is more accurate to provide a control action for a given function change under specific conditions which can be validated. This process requires the establishment of priorities with regard to which function changes are processed, and this involves the amount of deviation from the nominal value in relation to the accepted range of values. Functions that move outside their healthy range under normal conditions should have priority of correction to maintain a healthy system, even if not optimum: optimization will follow using the normal priorities. This steady-state procedure is overridden by fault conditions which are dealt with by a preselected sequence, when determined and identified by the protection system in conjunction with the PMS.

It is important, in small and island systems, to reduce control actions to the necessary minimum and to enable the system to recover from normal changes through its inherent dynamic response mechanism, by applying control to compensate for the primary reason for the condition change rather than by initiating several control actions based on a variety of function changes which, in themselves, might impair the system's inherent recovery process. Even under system fault conditions it is important to determine the nature of the fault and initiate the correct control actions to restore processes via the protection system.

During normal load changes there can be MW circulation between different generators as well as MW fluctuations resulting from inherent speed changes of the sets, but such transients should be rapidly attenuated by the AVR systems and governor controls; no external action at this stage is required, as both system voltage and frequency and generator MW and MVAR sharing will remain healthy and acceptable. Only when the steady consequent values are outside the control range will the PMS act to return the value slowly to within its operating range.

Specific situations, such as a satellite line where a change in load produces a voltage change which could be out of range, need rapid action to be initiated. In such instances, where the relationship between load change and voltage regulation is known, voltage adjustment can be initiated as soon as the load change signal is received. Alternatively, simple voltage control can be used to regulate the line voltage and receiving-end load voltage to within their operating range.

When a system has a complicated configuration it is even more essential to provide a simple control sequence to deal with the prime variables involved. In such a situation, if an attempt is made to adjust one function to be within its own operating range, its value will change again when another function is being dealt with. On the other hand, if an attempt is made to adjust all functions simultaneously, an unstable condition could easily be reached if adverse interaction should occur. The optimum procedure is to regulate the most extreme function towards its correct value, then repeat for the other functions which are outside range, in order of magnitude, and then recycle these functions until all are returned to within their operating range.

Thus, in a simple ring configuration, the voltage round the ring should be adjusted in steps, e.g. by the transformer tap-changer, until all the distribution voltages are brought within range, and then the sequence control can proceed to adjust MVAR circulation to bring the circuit MVA values within range. During this process the system voltages may require slight readjustment before the MVAR adjustment is finished.

On the assumption that the system components have been selected to be within their operating range for all normal working conditions, it will only be necessary to bring these values within the set values corresponding to optimum design operation, within a small margin.

In situations where the system conditions result in a serious overloading of some component, this condition should receive priority in the control sequence to keep it within design limits, if practicable. In general, however, this will only occur during abnormal conditions following maloperation, component failure or system fault, when instantaneous preprogrammed actions are initiated.

Interface functions

The factors relevant to ensuring high-integrity data, described above, also apply to PMS output control actions but many of the solutions are not practicable since an output control signal can usually have only one consequence which is to operate on some hardware control device. Unless there were direct feedback at this point, the PMS control system could only detect maloperation by some significant abnormal behaviour of the system, and by this time a critical situation could have arisen. In this respect, control logic is more critical than that applied to input signal data, as an erroneous signal could result in incorrect shutdown of essential plant at a critical time (possibly resulting in a total plant shutdown with consequent production and financial losses), or it might try to adjust some plant beyond its operational design limits.

When it is convenient to obtain a return signal indicating the immediate consequence of initiating a control action, e.g. the actual tap position on a tap-change transformer, it is a simple matter for the PMS to verify that its control signal has been correctly implemented. It is also possible to impose limits on the request for control action, since the permissable range of tap can be pre-established.

Where such direct consequent signals cannot be obtained for critical control functions it is necessary to utilize the nearest available, to try to minimize the consequences of erroneous signals.

Simple digital signals usually result in simple status change information which can be used to monitor action, but this has only a negative function in the event of a lost signal. With such signals it is possible to transmit duplex signals: if only one is received, action should be initiated and monitored. However, it is not always convenient to apply two alternative signals to a plant control interface without introducing a risk of common fault coupling. It is possible to get isolation by using a duplex mechanical interface, e.g. by using separate servo-systems, each with its own hydraulic system, but this extra cost and complication is only justified in very high integrity systems.

When the PMS control signal requires the provision of local services, such as electrical or hydraulic power, the provision of a healthy signal may not result in correct action, and can, in some instances, result in a dangerous situation. In such situations it is convenient for the PMS to send a warning signal that it intends to initiate control, and the local control system responds to indicate either that it is healthy and can respond, or that it is not capable of responding correctly. The PMS then sends its initiating signal, which performs the necessary control action or, in the latter case, it alarms that the control function is in default.

All the plant interface control components should meet the requirements given above for hardware. However, in many instances plant control equipment is not designed specifically for processor control and in some cases may be quite unsuitable. It would then be necessary to obtain a suitable control system which provides a compatible interface between the PMS and the component being controlled, although in some installations it may be possible to modify the existing control system. Such compromise interfaces often prove unsatisfactory, however, and should only be adopted if they have been found satisfactory in similar service on other installations. This is a particularly common problem where local controls have been developed from those provided originally for manual control by the addition of features to accept remote control signals of a conventional type, such as remote-power-type push-buttons.

Many local control interface equipments are designed exclusively for operation from a microprocessor, and several, such as governors, AVRs and protection devices, actually contain such devices; however, it is still necessary to ensure that the device is fully compatible with the PMS signal protocol before utilizing it for such a control mode. In extreme cases it may still be desirable to replace such plant control units with one designed specifically for operation from a microprocessor on a similar system which is functioning satisfactorily. The final decision on the choice of control equipment should only be made following a thorough reliability study incorporating the available alternatives.

System integrity

Analysis of integrity

The optimum choice of control action to be taken following any system change, whether normal or abnormal, cannot be determined arbitrarily. It requires a detailed and organized study of the complete power system, including all its control systems, as well as the PMS control unit.

With such a total system it is necessary to determine not only the local effects of failure of any component, but also the consequent interactive actions. It is equally important to determine the level of maintainability of all, or at least of the essential, components of the system, since some failures will result in loss of parts of the system, and the nature of necessary repair or replacement is just as significant as the reliability of each component.

Such a study is very extensive and can be time-consuming if a satisfactory procedure is not adopted and adhered to. It is now customary to adopt one of the standard procedures available: this can ensure that all aspects are dealt with adequately, and the procedure can be used by several operatives dealing with different parts of the system, on a common basis which can be integrated to form the complete study.

Failure modes and effects analysis

The purpose of such a study is to investigate the consequences of all component failures on the total system and to estimate their relative importance. This can be done by examining individual hardware items or alternatively by assessing the effects of function or signal failure, or possibly by a combination of these. It is also essential to identify the means by which the failure can be detected. A comprehensive analysis would deal with every item in a system. In practice, it is usual to consider some integrated items. For example, an AVR can be regarded as a unit and its functional behaviour will depend on component failure or maloperation: these latter need not be considered in detail. The level of analysis must be accurately identified, usually in the form of block diagrams, and by using a standardized nomenclature these can easily be cross-referenced to simplify interactive effect evaluation.

When the effect of a single failure may be undetected it is necessary to extend the study to include the effects of all relevant second failures and treat these in the same way as primary failures.

The existence of redundancy, standby plant and autotransfer systems should be incorporated in a failure mode analysis, as this is very relevant when applying a severity classification to the failures to determine some order of priority of significance varying from catastrophic down to negligible.

In electrical power studies it is also essential to include severity factors in relation to loads, drives and distribution units, with regard to their importance to the purposes for which the system is installed. For completeness, this necessitates knowledge of the transient behaviour of the system following the appropriate failures being analysed.

When the complete failure mode and effects analysis has been completed, an analysis should be made of the results to determine what changes would be required to the design to reduce the failures producing the most severe results, particularly if these are classed as catastrophic. Usually it is found that the severity classification tends to divide failures into groups. When the most severe have been eliminated, a stage will be reached at which it will be impractical or uneconomic to try to eliminate the next most severe group. Redesign should be restricted to the higher classification failure modes.

At this stage, a criticality analysis should be carried out which incorporates a probability of occurrence for each considered failure mode. This, together with the severity classification already established, indicates the relative reliability or criticality of each failure mode. The severe consequences of a failure mode can be offset by its improbability of occurrence, and this affects the relative standing of the individual items in the severity classification.

Maintainability

The final significant factor to be included in the study is the maintainability of the components of the system. This includes such basic features as redundancy, where the failure of a component is immediately compensated for by a replacement or alternative system which leaves the electrical system stable and continuing to operate normally. It should be noted that where redundancy is provided, the MTBF of a device is not so significant and will not then appear high in the criticality analysis.

Other factors can maintain system stability within lower operational levels in the event of a particular failure mode and the level of acceptability of such reduced operation can also be included in the study.

The actual maintainability of a component or unit is a measure of the importance of the device and the amount of time and complication involved in either replacing or repairing the failed unit. This is often related to the above reduced function mode; for example, when a generator AVR fails, if automatic transfer to standby manual operation is provided, the system will remain stable and provide normal power requirements. The availability of a spare AVR will enable the faulty unit to be replaced in a specific time either with shutdown of the faulty set or, if designed appropriately, by an on-line replacement. Should adequate generation capacity be available then the consequences of an AVR failure will be low on the criticality analysis, provided that the necessary features have been included in the plant design.

A reliability analysis carried out on the above principles can indicate the real criticality of all failure modes and highlight those items or aspects which should be improved. The cost of modification or change to design to produce the same result must, of course, be assessed in relation to the possible consequences of the particular failure mode.

Safety-related systems

In addition to the integrity of the PMS and the electrical generation system, there are frequently systems installed to improve the safety of installations, fed from the supply and able to interact with both the PMS and the electrical system. Most processes supplied with the electricity generated have their own integral monitoring system, which is arranged to indicate dangerous or abnormal conditions, and this may be arranged to initiate corrective or emergency actions. In many instances such action will involve the electrical supply at some level.

When generation plant is dedicated to a local process or plant, the interaction is very close and it is logical to interconnect any plant safety or emergency system with the PMS, which can coordinate any generation modification required by conditions arising within the plant. This condition arises in offshore oil production platforms and on board ships, and in both situations the safety of the plant directly involves the safety of the operating personnel as well as the supporting structure itself. In the past, separate emergency shutdown systems and fire and gas systems were employed to initiate necessary actions in the event of any serious emergency, but the standards which these were required to meet were not always satisfactory. In view of recent incidents, the importance of such systems has been reassessed; they are now classified as **safety-critical systems**, as opposed to safety-related systems, which merely may affect safety.

Considerable progress has been made in the production of standards, both national and international, to establish a satisfactory common basis for all such systems to try to obtain a more realistic and satisfactory approach to their design. This, of course, can extend into the design of PMS units for generation plant associated with such systems.

When such conditions are likely to arise it is essential to determine what the overriding safety criteria will be, as it may be preferable to let some items of plant run to destruction if, by so doing, the safety of the rest of the system, or the personnel, can be obtained. Thus the emergency generator on a ship in distress must be maintained as long as possible to provide power for SOS communications, emergency lighting and the operation of life-saving equipment.

5

Energy management systems

The power management systems (PMS) discussed previously have been intended to control a power system to provide stability and continuity of supply up to the required design conditions of loading and to ensure that the plant involved does not exceed its rated conditions. This assumes that the plant components have been designed for all the operating conditions which will arise and that the necessary criteria have been incorporated in the PMS databank. Operating efficiency and economic factors have not specifically been discussed, on the assumption that these have been dealt with when specifying the plant involved and the operating conditions to be met. Should the control system be required to deal with aspects of efficiency or economy, it is necessary to consider the total energy involvement in the system. This can be done by a natural extension of the PMS, incorporating all the features as already described, as required, but including the necessary data to determine the utilization of total system energy in relation to the level of operation considered necessary for or relevant to acceptable system operation.

As in the case of a PMS, an energy management system (EMS) can be used in two ways:

1 To assist the operator by monitoring and presenting data, including cost if necessary, by analysing this and advising actions to be taken to meet the criteria included in the predetermined data supplied to it.
2 To perform the above and, in addition, to implement automatic control of the system plant, such as generator sets and other

energy sources, in accordance with its deduced analyses and conclusions. It will, of course, still function as a PMS to keep plant within designed operating limits, but such limits may have to differ from those used if it were performing solely as a PMS and not as an EMS.

Energy analyses of systems are nearly always associated with costs; in some instances the analysis is used solely to compare performances of alternative packages, e.g. by assessing the overall efficiency of generating sets, as this is a simpler proposition. Ultimately, of course, the study will extend to an economic assessment of a system or subsystem, but this need only be done when some of the alternative options of components have been eliminated by the simpler efficiency approach applied to the total generation package, e.g. a combined generation set incorporating waste heat recovery.

The assessment of costs can be performed approximately by examining only the electrical energy used in a power system. However, where optional sources are available, such as imported power from another system, or locally generated power, it is necessary to estimate the two separately and assess the total cost on a common basis to determine if the conditions selected are optimum, or whether some relative adjustment would be more economical.

The determination of the criteria to be used for any such relative assessment can also be very simply based, e.g. on fixed MW hour rates, or may be very complicated, incorporating all relative cost factors in the system. The level of complexity must be determined by the importance of the system and the benefits which could accrue from the incorporation of an efficient EMS. Thus it may be necessary to utilize a detailed study of some components, such as combined generation sets, but to use a lower degree of complication for the rest of the system involved.

Electrical energy

A supply of electrical power, expressed in MW, for a specific time, corresponds to an energy value which can be expressed in MW hours. This value can be metered conveniently and is normally used as the basis for determining energy demand and hence cost. A rate per MW hour does not necessarily relate to real cost of producing the energy. When using this simple basis for costing power, it is necessary to apply limiting conditions to its use and these are included in supply tariffs used for costing electrical energy consumption.

Most devices which distribute or consume electrical energy also require an MVAR supply, and the real cost of supplying MW power is related to the power factor at which it is supplied. In the interest of simplicity this parameter is not included in normal costing but it is usual to impose a limit on the demand PF or demanded MVAR above which extra charges will be applied.

There is, of course, a limit on the supply capability imposed by the distribution system cables, transformers etc. and the demand must not exceed this value in normal service. Any increase in demand above this limit will require considerable capital cost expenditure to increase the system capacity and could involve a significant time delay in being implemented.

The availability of local generation can provide an alternative to increasing incoming power supply capacity. A local generator is more usually installed to provide an increase in integrity of supply, either by functioning as a back-up to the existing incoming supply, or by being used as the main source of supply, with the incoming supply as the back-up. In many instances the two sources of supply are used in parallel and the distribution of load is determined by the control system provided.

The size of set installed in relation to the total plant demand is determined by several factors, as discussed previously; the nature of the load, e.g. continuous process or intermittent, peaky or relatively steady, and the magnitude of any local 'essential' load are usually the most significant factors. In addition to determining the capacity of the local set, they also determine the optimum control procedure to be adapted for the plant power management.

The cost of electrical energy produced by a local generator set will not be the same as that provided by the supply authority and cannot easily be directly related to it. Thus, even if a MW hour cost is derived related to fuel costs only, this will only be correct at one load condition. It will vary with the actual set loading, and cannot be used as a directly comparable cost. To obtain a reasonably comparable cost of locally produced energy it is necessary to take into account, as well as fuel cost, capital investment costs, maintenance and service costs, and costs resulting from any abnormal or emergency operating conditions. Some of these factors can be evaluated reasonably accurately but some will involve statistical analysis based on experience, as available. In practice, some of the advantages, or disadvantages, associated with local generation may have associated costs related to effects on the local plant; it may be necessary or desirable to make allowances for these when determining the value of local generated energy for comparison purposes. Thus, when the installation of local generation is economically justified by the increased reliability of supply, and hence integrity of the process output, it is quite reasonable to compare imported electrical energy with local generation costs on the sole basis of fuel costs plus a factor to compensate for maintenance costs and possibly deterioration or replacement costs.

With local generation it is easy to control MVAR generation by adjusting excitation. Variation of MVAR does not have a significant effect on the generator efficiency, but its effect on the generator MVA can affect the winding heating and should be kept within the generator rated value. The most energy-efficient mode of operation is to operate the generator at unity PF, i.e. providing no MVAR, as this corresponds to minimum excitation and MVA and hence minimum heating and losses. Under this condition all the local load MVAR demand would have to be supplied from the incoming power supply irrespective of any imported MW, but this would not affect the MW hour consumption. However, should there be some PF or MVAR limit imposed by the power tariff, this could involve extra charges, and in such conditions it would be beneficial to produce local MVAR to avoid entering the penalty region in the imported supply.

It is advisable, therefore, to organize the local MVAR control to optimize the charges on the incoming power up to the safe loading on the local generator. If no financial benefit accrues from operating the generator up to its maximum available MVAR capability, by nature of the tariff existing, then there is no advantage in using such a control logic.

When more than one local generator is in service, each should have its own power-costing determined as they could be significantly different. In the event of multiple sets being kept on the bars to increase supply integrity, e.g. through redundancy, the optimum loading of each set should be determined: for example, it may be beneficial to run some sets on base load, and use others for peak loads where their part load performances are superior to those of the base-load sets.

With multiple local generation the principles described above can be applied to the integrated locally generated MVAR irrespective of the relative MW loadings being used, as long as the same overall criteria are applied to the individual sets and the total generated values.

When alternative fuels are available for local generation these should be separately costed and related to the appropriate power cost formulae for each set, so permitting valid optimization of total cost. However, some overriding factors may have to be imposed; for example, some fuels may only be held as a short-term stock for use in special situations or emergencies, while others may have an unacceptable effect on the environment and their use may have to be restricted to limited periods of time.

Thermal energy

The electrical energy produced by a generator is derived from the mechanical output from its prime mover which, in the majority of instances, utilizes energy in some other form. Such energy conversion involves losses, and the overall efficiency of conversion is a significant factor in the choice of prime mover for any particular generator.

The most commonly used prime movers utilize thermal energy obtained from the combustion of fuel either directly in the engine or in a boiler producing steam for use in the prime mover. The former group comprises direct fuel combustion engines such as petrol oil or gas engines and gas turbines, while the latter group includes steam engines and turbines. Steam engines and turbines are commonly used in applications where steam is required for process purposes and it is advantageous to use some of the energy to provide the electrical power required by the process. It is often convenient to use a steam turbine as a back-pressure set by taking steam at boiler pressure and exhausting it at a reduced pressure which is still usable in the process: that is, it acts as a pressure-reducing device which also generates electricity.

In general, the majority of generator sets are driven by some form of combustion engine or turbine using some form of primary fuel. There is a wide range of possible devices, each with advantages and disadvantages which vary in significance with the size and application of the unit. In all areas, however, the overall energy conversion efficiency is a significant factor in the choice of engine for a prime mover. This simple efficiency value varies over wide limits depending on type, speed and size of generator set. In general, plant with higher efficiency also has a higher first cost, and the relation of running costs to capital costs can be a critical factor in the choice between alternatives which are otherwise technically equivalent. Energy loss in the conversion process appears mainly in the form of thermal energy, which can be in a variety of forms, depending on the type of prime mover, but can be classified as high-grade or low-grade heat depending on the temperature at which it is produced which, in turn, determines the use to which it could possibly be put. If any of the waste heat can be utilized, then the nominal energy conversion efficiency of the generator set can be increased and show an economic advantage over similar sets which do not utilize such thermal energy. Of course, the utilization of such waste heat invariably involves extra expenditure, which must be offset against the improved efficiency. In some instances the utilization of waste heat may reduce secondary problems associated with handling or removing it, and this may result in some saving on plant costs.

The largest source of waste heat is in the products of combustion; exhaust gases which are at a high temperature can be a convenient source of recoverable waste heat. They are usually suitable for raising steam in an appropriate heat exchanger or for heating other fluids required in the process.

Other energy sources

Electrical generators can also be conveniently driven by turbines utilizing gravity or pressure head in water, fluid or gas systems. Hydraulic turbines are commonly used to drive electrical generators where an adequate flow of water is available with sufficient head drop. Small units can prove very economic where the hydraulic flow system already exists, and the main cost of installation is the turbine generator set itself. Where considerable civil works are required to contain and control the water flow, the capital cost can be considerable. It is usually only justified when large sets can be installed with considerable electrical power output.

Another convenient source of power to drive a turbine is in a fluid or gas flow line, where the mass flow is such that a small pressure drop across the turbine does not interfere with the purpose of the fluid flow. Such turbines are frequently installed in the exhaust flow of large oil engines, where they utilize waste energy to drive a supercharger, which boosts the air flow to the engine to improve its combustion capability. These units result in an improved overall efficiency but can also increase the actual output for a given size of engine.

Similar types of turbine can be incorporated in pipelines and so produce an output usable by an electrical generator to produce a local power supply. This can be an economic arrangement in remote locations where the cost of providing an electrical distribution system would be excessive in relation to the relatively low demand. This arrangement is also used where line pressure reduction is required for control purposes, as the electrical energy produced is related to the pressure drop across the turbine, the flow remaining unchanged. This system provides a simple pressure regulator in which the pressure energy dropped is converted into electrical power which can be used locally or fed back into a local system.

With such generator sets it is not appropriate to try to cost the electrical energy on the basis of energy costs. An economic balance must be performed by relating the cost of the installed generation plant to the savings obtained by avoiding the input of electrical energy from an external power supply, or to the revenue from the sale of the power generated. By nature of the locations in which such generation plant is installed, the installation cost of an electrical supply from an external system is usually very high when compared with the actual cost of the power required.

Hydraulically driven generators are similar in that it is difficult to allocate a unit cost to electricity produced, since the real cost is related to the capital cost of the installation together with associated financial charges and becomes an accounting procedure rather than purely an energy cost.

Prime mover fuels

The technical factors involved in choosing the fuel for a generator prime mover have already been considered. When making an economic comparison, it is necessary to consider additional aspects such as cost of the fuel handling, the effects of fuel choice on the reliability of the system and also its effects on the environment.

The integrity of an electrical generator's output is directly dependent on the health of its prime mover, which in turn requires an adequate supply of suitable fuel at the specified design conditions. Some fuels, such as gas, may be obtained from an external mains supply and this has a distinct advantage in providing a continuous supply. However, this supply is outside the control of the consumer, who must rely on the supplier to provide an uninterruptible and adequate supply. The conditions of supply must be satisfactory, in that the quality of fuel must be within limits acceptable to the prime mover and the pressure available must meet peak flow requirements. Deficiency in either of these conditions can reduce the available generator output and, in the extreme case, can eliminate it. It is difficult for the supplier to guarantee to meet design conditions without any interruption; contractual agreements can be made to cover any default, but its significance for the integrity of the electrical system must be carefully assessed.

Some fuels, such as oil, can be stored conveniently in suitable tanks or containers, and this avoids dependence on a continuous external supply. However, replenishment is necessary, and the magnitude of the amount stored must meet the maximum demand between guaranteed delivery intervals. Non-availability of replenishment with fuel of acceptable quality can result in loss of output from a generator set.

Reliability of fuel supply can obviously be increased by increasing the amount held in store, but this can involve considerable increase in capital and running costs which will be a significant factor when relating reliability to cost.

One solution which can relieve the above problems is to use both systems: continuous incoming supply together with an adequate local storage facility. However, while a gas supply is normally available from a centralized source, local storage can be bulky and expensive. Conversely, oil can be conveniently stored locally but is not commonly available from a mains supply system.

Where it is possible to use a dual-fuel prime mover, such as a gas turbine which can also operate on an oil supply, it is quite practicable to operate it normally on gas from an external mains supply, and to provide an adequate local store of oil to cover the non-availability period of a gas supply shutdown. When suitable fuels are available it is possible to provide automatic changeover from one fuel to the other without significant disturbance to the electrical generator output, using particular types of prime mover provided with suitably designed governors and fuel systems.

When considering the economics of prime mover fuels it is necessary to establish what arrangements are being made to meet the required reliability of the generator on the system and to make due allowance for these.

Fuel handling can involve considerable expense. Particularly where local storage is being provided, the requirements of the associated hazardous areas can involve the use of certified equipment and the application of special safety precautions to meet standards acceptable to the appropriate approving authority. The necessity for overhaul, maintenance and cleaning of fuel systems can also introduce a significant overhead cost, and the extra precautions required to avoid the contamination of fuel may, in some cases, involve an expensive operation.

The actual quality of the fuel is directly relevant to satisfactory operation. Standard specifications do not always limit levels of trace elements, such as sodium, which can produce damaging conditions on gas turbine blades; where contamination is known to exist or is suspected, filtering equipment may have to be provided. Gas turbines consume large quantities of air to meet correct fuel combustion requirements, and in a marine environment this can result in ingestion of sodium salts, which can have similar deleterious effects on the blades. It may be necessary to provide suitable filters to reduce such contamination.

Fuel quality is also significant with regard to prime mover exhaust emissions and their possible effects on the environment, which can also be affected by the mode of operation of the plant.

In some environments restrictions are placed on the use of certain industrial effluents and this can prove a serious factor in determining the most suitable fuel for a generator set. Dealing with effluent to meet the required standards can involve considerable extra cost, which should be included in any capital cost assessment of a projected generation plant.

Boiler fuels

All the factors described for prime mover fuels apply to fuels used in boilers raising steam for generator prime movers, but to reach an economic comparison in relation to electrical power generated involves much more complication, and the number of possible alternative options can be quite high. It is necessary to replace the simple combination of fuel plus prime mover, and the incorporation of a steam loop gives another set of optional parameters relating to the steam conditions and hence results in a much wider choice of prime mover type. However, in most installations of this type use is made of produced steam for other purposes such as heating or for use in an industrial process; the value of this service becomes a factor in the overall economics, and in many instances can become the main factor.

In installations where the value of produced steam is more significant than that used for electrical energy production, the electrical generating set will be selected to make the best use of whatever steam conditions are being made available for this purpose. This will reduce the number of viable alternative generation packages. It is preferable, of course, to provide 'standard' steam conditions that match 'standard' generator prime movers, if at all possible, as the cost of obtaining special machines for unusual conditions could be excessive and add complication. This should be weighed against possible problems in providing the 'generation' steam at preferred conditions.

Evolution of CHP

Although the term CHP (combined heat and power) has come into widespread use only in recent years, the concept of using the dual energy outputs resulting from the production of electricity from fossil fuel has been implemented since the beginning of this century because the effective increase in overall conversion efficiency resulted in financial savings. The most common systems involved the utilization of some of the waste heat from a conventional generating set, e.g. by passing the products of combustion through some form of heat exchanger to obtain a source of hot water for industrial or domestic heating. When high-grade thermal energy was required, this could be obtained by increasing the quantity of primary fuel in a boiler to raise steam, both to drive the electrical generator prime mover and to provide suitable steam for process purposes or direct heating. This latter type of scheme increased the initial capital cost above that required for a simple generator installation, but was probably lower than the total cost of plant to produce the two forms of energy independently. Operating costs were advantageous, as the overall energy conversion from the primary fuel could be appreciably higher using CHP; this resulted in a much more economical installation. Several arrangements of steam-driven prime movers were used in such schemes, including both condensing turbines, in which the full boiler pressure was utilized, and back-pressure turbines, in which the turbine accepted boiler pressure steam and reduced this to a pressure more directly usable by the process. This enabled higher boiler pressure to be adopted, resulting in higher boiler efficiencies.

Elaboration of these arrangements could be provided by bleeding off steam at intermediate pressures from the turbine to meet special process requirements, as required, or utilizing different turbine stages, e.g. combined high-pressure and low-pressure turbines driving a single generator.

In the UK this concept tended to increase in application, since it matched the requirements of both domestic and industrial users for locally produced power in relatively small amounts. As larger industrial units evolved, usually being consumers of large amounts of thermal energy as well as electricity, the same principles were still applicable and private generation became quite common, usually incorporating some form of CHP.

However, as the amount of electrical power consumed increased, legislation was introduced to organize and control its use. The Acts of 1899, 1909 and 1919 placed restrictions on private generation. As a result, the supply industry gradually became a virtual monopoly as, in addition to meeting the legal requirements, any prospective generator had to obtain prior approval from the Electricity Commissioner. In 1937 the Electricity Supply Regulations laid down the technical requirements but, as these were not sufficiently comprehensive and lacked detail on specific aspects, they were supplemented by a series of documents issued by the Electricity Supply Industry itself. These were enforced equally rigidly, being regarded as virtual extensions of the legal requirements. This situation remained until 1988. In 1947, the Electricity Act nationalized the electrical industry. The pattern of industry started to change significantly in the 1950s and 1960s and the demand grew for cheaper electricity, without the simultaneous need for thermal energy. This led to the development of a large-scale generation and distribution policy. To increase generation efficiency, larger generating sets and stations were designed and built with efficiency values approaching 40 per cent. These large stations, however, required a build time of 6–10 years; surplus generating capacity was required to cope with prospective growth of demand over such extended periods, and this reached a value of 20 per cent. Transmission efficiencies were improved by installing a national grid system, which utilized higher voltages necessitating more and larger pylons.

Public interest in the environment started to become a significant force. Objections were raised to the appearance of large cooling towers and lines of pylons traversing the countryside, to the detriment of its beauty. Other objections concerned the heating of rivers, to the detriment of aquatic life, by the cooling water from the condensation of steam from the turbines, and the unsightly and often messy procedures adopted to dispose of boiler ash. Finally, the effects of chimney emission of objectionable products, including carbon dioxide, sulphur and NO_x gases, became a matter of worldwide concern as their association with 'acid rain' damage was gradually established; means of reducing dangerous emissions were investigated, but were found to be quite complicated and costly.

Another subject of worldwide interest was that of energy conservation and the fact that the large generating stations, while providing up to 40 per cent energy usefully converted to electricity, also wasted the residual 60 per cent. The combination of these 'environmental' factors, and other factors associated with changes in the industrial basis of the UK, led to two lines of investigation: to improve current generation policy to meet the various criticisms; and to consider alternative means of satisfying energy demands. These obviously required a review of the existing statutory situation and its modernization to meet the changed situation and requirements for the future.

Alternative energy sources were investigated for a variety of reasons: to reduce adverse effects on the environment; to reduce the consumption of irreplaceable energy fuel; and to raise conversion efficiency by the use of alternative processes or plant. Nuclear reactors were developed which purported to satisfy some of these requirements, but a popular antinuclear movement, largely activated in relation to defence policy, proved a severe handicap. Moreover, when the hazards associated with nuclear plants were fully investigated, and safety precautions implemented which were regarded as satisfactory the costs of such stations increased to such an extent that there is doubt as to their future role in electrical generation.

Other forms of energy, such as hydro, tidal, solar and wind power, were shown to be viable although limited in the number of practicable sites in the UK and hence in the amount of energy available from such sources. However, even these forms of energy utilization have attracted the attention of the environmentalists and they have had little impact on the total energy market.

Other novel sources of energy have been investigated which, by using undesirable or waste products, have the support of environmentalists. Prime movers can be fuelled by sewage gas or methane obtained from old landfill sites; boilers can be fired with suitable household or industrial waste and the steam used to provide electricity if required. Such schemes are now operating satisfactorily on a relatively small scale.

More sophisticated sources of energy have been researched but, while some have proved practicable, the investment and time needed to develop them to production level preclude their becoming widely available in the near future.

At about the same time as these developments, the petrocarbon discoveries in the North Sea brought a surge of interest in the prospect of a new local source of cheap energy: a convenient fuel with greatly reduced levels of sulphur and carbon dioxide produced from its combustion, as compared with the commonly used alternative fossil fuel, coal. This resulted in a surge of development in the use of gas turbines which utilized a lot of the technology resulting from World War II experience gained on propulsion equipment for both ships and aircraft, and since continued for industrial purposes. The development of steam injection to gas turbines was found to reduce the NO_x emission significantly, increase the power output from the turbine, and reduce the heat-to-power ratio to a value attractive to some CHP schemes.

North Sea prospecting continued and, as the scope of the resources was gradually appreciated, the provision of oil and gas to consumers resulted in pipelines from offshore sites to shore and also in landlines to distribute the products to prospective consumer locations. The advantages of the gas, as compared with existing locally produced gas, immediately changed the relative market values of alternative fuels, and the availability of high-pressure gas throughout the country changed the economics of fuel provision for industrial users. However, public opinion again influenced industrial attitudes. The subject of energy conservation and the demand for a government energy policy created a situation in which the efficiency of energy conversion became a major consideration.

At about this time, other factors started to affect the mode of using energy. One of the most significant was the growing need to replace boiler and other auxiliary plant that was reaching the end of its useful life. When replacement costs were being assessed, the new concept of CHP was often included as a viable alternative to a straight replacement of the original boiler plant by installing a steam-driven or gas-turbine-driven generator, particularly where the heat-to-power ratio required matched that provided by the proposed CHP plant. A similar situation was arising in the electrical distribution system, where many local areas had reached their limits of safe loading as a result of the continual growth of demand since their installation. The need to provide further additional power could only be met by the installation of new equipment or by modification or uprating of existing equipment, which could involve considerable extra cost to the prospective consumer. This single factor quite often renders a proposed CHP scheme non-viable because of the extra cost involved.

As attempts were made to promote CHP schemes locally, conflict with the existing electrical supply industry philosophy, combined with the difficulties and restrictions arising from the existing mass of legislation and regulating documents, presented great problems and resulted in a situation where the installed capacity of CHP in the UK compared very unfavourably with the amount in use overseas. For example, in one year total orders for gas-turbine-based CHP installations were 25 times greater in the rest of Europe than in the UK, and 800 times greater in North America. The amount of CHP installed as a percentage of the total generation capacity is about 60 per cent in Denmark, 25 per cent in Holland and about 4 per cent in the UK, and even the target declared by the UK government to be reached by the year 2000 is only 4 GW, which is about double the present amount in this country.

As a result of interest aroused in the government, in 1978 the Energy Efficiency Office of the Department of Industry, supported by the Energy Technology Support Unit and the Building Research Energy Conversion Support Unit, initiated the Energy Efficiency Research and Development and Energy Efficiency Demonstration Schemes, and several hundred projects were implemented. These promotions were replaced in 1989 by Best Practice Programmes; Good Practice Case Studies supplemented by Good Practice Guides provided information on established, economically viable energy efficiency measures, while Energy Consumption Guides provided information on energy uses in particular installations. At the same time attempts were made to remove restrictive legislative and institutionalized barriers to the use of CHP.

The situation started to change in the 1980s and in 1983 the Energy Act enabled electricity to be generated by persons other than electricity boards; the 1984 Regulations identified the information that a prospective generator had to provide to the electricity board under the 1983 Act. In 1988, the Electricity Supply Regulations replaced the 1937 Act and, finally, in 1989 the Electricity Act repealed the Acts of 1899, 1909, 1919, 1947 and most of the 1983 Act and brought about the privatization of the industry.

However, rather than clarifying the situation as far as prospective CHP users were concerned, the reorganization of the industry introduced new complications. The new electricity supply industry is divided into several main generators, a national grid company (NGC) and twelve regional electricity companies (REC) operating a 'pooling and settlement' system in which all generators sell to the pool and all suppliers are required to buy from it. The pool price is made up of five components: energy, 65 per cent; distribution (REC), 15 per cent; transmission (NGC), 5 per cent; a fossil fuel levy to assist the nuclear generation industry, 11 per cent; and the remaining 4 per cent from miscellaneous charges. This pool price is updated half-hourly and can fluctuate widely and rapidly; for example, in one 24-hour period the ratio of maximum to minimum can be as much as 3:1, and there may be five or six cycles during this period. If a supplier were to offer a fixed price contract to a large consumer it would be usual to cover such fluctuations by entering into a contract for differences with a generator which would provide a give/take payment related to deviations from the nominal; however, the Office of Electricity Regulation applies price-capping which imposes a measure of restraint.

All users of 1 MW or over are free to buy from any REC and there is a wide range of alternative deals which range from a simple fixed rate through to paying a straight pool price, with several hybrid arrangements between these. Alternative forms include two-rate day/night tariffs, seasonal time of day, maximum demand and load management schemes of payment. Users of less than 1 MW have an entirely separate pricing arrangement.

As a consequence of the new Act, just as the national power and power generation companies can also act as suppliers, the RECs can also become involved in generation. In addition, various large consortia are interested in generation and propose building large stations for this purpose. Installations above 50 MW can trade through the pool, those above 100 MW being centrally despatched by the NGC. The criteria involved in these new proposed designs vary widely and some of the advantages of CHP render this a very attractive option for these new projects, as the overall efficiency of a CHP installation can be as high as 85 per cent, as compared with about 80 per cent for a simple boiler system and a value of about 35 per cent for a simple electrical generation set.

CHP schemes can be conveniently separated into three distinct groups, each with its own unique problems and operating requirements. Large units above about 50 MW are essentially generating stations designed to utilize their waste heat efficiently and form an effective part of the electrical system. Their interconnection with the system must be fully planned and developed well in advance of installation. They usually consist of completely new equipment and plant, and their design is carried out as a single composite project, being self-contained apart from the system interconnection.

CHP schemes less than about 1 MW, usually supplied in the KW range, have negligible effects on the electrical power supply system and are often standardized package units with about 75 per cent overall efficiency, utilizing a spark-ignition engine prime mover. The generator is operated in parallel with the public supply to reduce the site demand, and the engine waste heat can be used for local heating. A high-efficiency gas boiler may be included as a back-up for heating in the event of the set being shut down.

The third group is in the intermediate MW range. Their capacity can produce significant effects on the power supply system and they are usually installed at an existing site which has a fully developed electricity distribution system, often with restricted capability for increase in capacity or modification. These installations are typically found in process industries, such as chemical, paper or food production plant, which have a considerable demand for steam as well as for electrical power and may involve continuous processes requiring high integrity of energy supply. The ratio of thermal to electrical power demands depends upon the

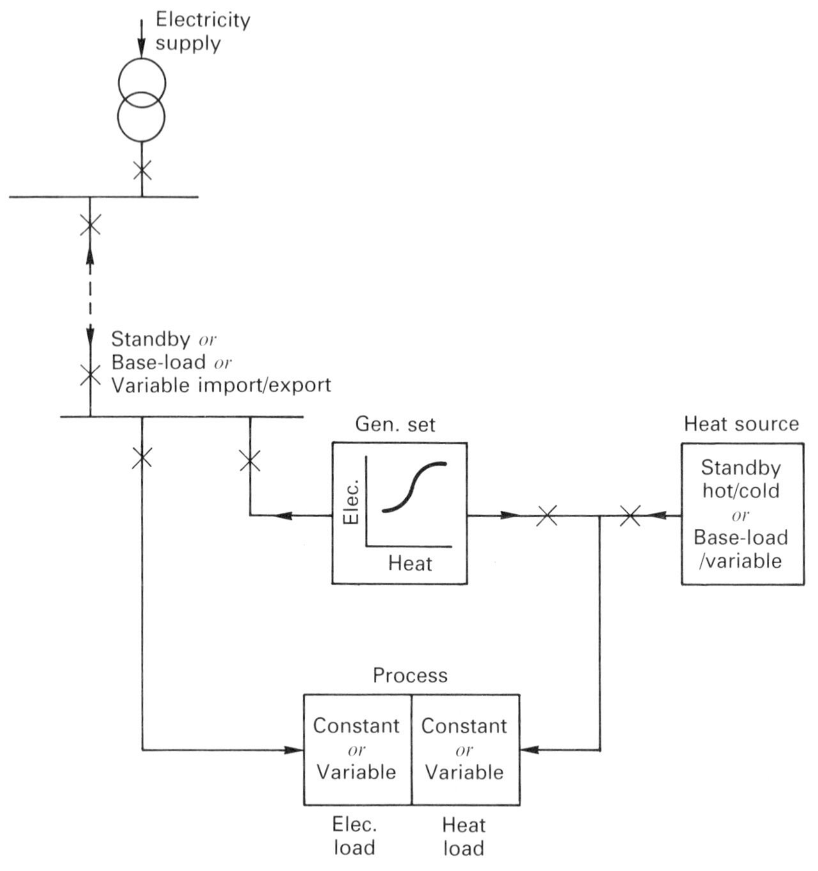

Figure 5.1 *CHP simple block diagram*

actual processes involved and changes with modification to the process or growth requirements. It can also vary significantly with time of day, week or year, and it is consequently difficult to provide the actual energy required economically.

As this group of CHP systems manifests all the problems, it can be used to analyse these in detail. The conclusions will be applicable to all other CHP schemes, as relevant. Figure 5.1 shows a simplified block diagram for a typical CHP installation.

Advantages of CHP schemes

1 Improved conversion efficiency can save energy and money.
2 Can improve security of supply.

3 Can be environmentally friendly.
4 Can use alternative fuels
 (a) high-pressure gas from line.
 (b) liquid or solid fuel.
 (c) waste products.
5 Available technology meets requirements and can give energy control.
6 Reduced electrical supply demand.
7 Can benefit from available costs when replacing old plant.

Disadvantages of CHP schemes

1 Requires investment in non-core business.
2 Requires financial funding.
3 High capital cost, payback time possibly 3–5 years.
4 Uncertainties
 (a) energy costs.
 (b) environmental legislation e.g. Carbon Tax.
 (c) future energy requirements.
 (d) reliability of equipment.

Barriers to CHP schemes

1 Agreement, regulations, codes etc. to be complied with.
2 Licensing of generators.
3 Attitudes of some RECs.
4 Interface with pool and trading of surplus electricity.

An inadequate study of all the considerations listed above, prior to deciding whether to install a CHP system, can result in the adoption of a solution unsuitable to the conditions relevant to any particular site; this in turn can result in financial loss, and possibly future technical problems with the plant. In order to get full benefit from the possible advantages and remove as many barriers and disadvantages as possible, it is essential to utilize the services of engineers or advisors trained and experienced in the special problems associated with this unique form of generation equipment. This can be ensured by arranging for suitable assistance to existing plant personnel, if they do not have much experience, or by employing a specialist company to evaluate feasibility and prepare specifications for a suitable CHP installation. These specialists can also be used to procure, install, commission and put to work the plant and, if required, they can provide training or operational assistance for a period to enable the regular operators to gain confidence in handling the new installation.

As a result of the growing interest in CHP schemes, and because of the growing size and complexity of proposals being considered, specialist companies are now being formed with whom it is possible for a prospective user to enter into some form of liaison agreement such as a joint venture, contract energy management scheme, or heat and power purchase scheme. With the last-mentioned scheme the consumer is charged for the electricity produced at a rate which incorporates capital, installation and maintenance charges and also fuel consumed. The consumer benefits from waste heat recovery from the prime mover, and gains lower total energy costs without initial capital expenditure.

Under a contract energy management scheme the contractor supplies the necessary capital, carries out the feasibility study, designs, builds and installs the necessary plant and, at least initially, operates it and demonstrates that it meets the guaranteed efficiency of performance. The consumer makes regular agreed capital repayments, pays for fuel consumed and operates the plant which, at the end of an agreed period, say 10 years, will revert to the user.

There are many possible permutations of such arrangements, which are basically intended to remove the disadvantages as far as possible.

Interconnection of CHP and electricity supply systems

When a CHP scheme is being connected to an incoming electricity supply, operating conditions are very different from those on connection of a simple electrical load. The following aspects are unique and require careful consideration to ensure that a satisfactory, safe and acceptable installation results.

1 Increased fault level on system.
2 Isolation requirements.
3 Protection discrimination.
4 Effects of loss of supply.
5 Effects of power export/import on supply.
6 Metering requirements on supply.

7 Requirements of control when alternative island mode is required.

Increased fault level

Distribution systems are normally designed to deal with the fault level contribution from the supply system itself, plus a usually smaller in-feed from site rotating plant, with some allowance for future extensions or system changes. It is desirable, therefore, to limit any changes to meet the existing capability and avoid having to change the distribution equipment, as this could be expensive and time-consuming. Incoming switchgear on a consumer's site is usually matched to the supply authority's incoming switchgear, and frequently forms part of a combined switchboard capable of dealing with a site fault fed from the incoming supply. The addition of new generation on the site increases the total site fault contribution, depending upon its subtransient reactance, and this total is usually within the capacity of the consumer switchgear. When it is added to the incoming contribution from the supply system, however, the total may exceed existing switchgear nominal rating for some faults on the site. In this case, unless the existing switchgear is to be changed, or can be uprated adequately, the new additional contribution must be reduced. This can be done by increasing the generator reactance, adding equipment to increase the system reactance, or reconfiguring the system. The appropriate reactance of induction generators, or motors, is dependent upon the starting current and torque values, and can usually be adjusted to a more suitable value at the design stage of new equipment. With synchronous generators, however, significant change to the subtransient reactance can increase its size and cost, and may also have adverse effects on its transient reactance and hence on the transient stability of the system.

The incorporation of series reactors in the local system reduces the fault contribution, but the voltage drop across the reactor results in adverse 'voltage regulation' problems on the site distribution system as a result of normal load flow through the reactor.

Resonant links can effectively reduce through-fault contribution to less than twice the normal throughput load, as determined by its situation in the system, and do not present a serious voltage regulation problem. However such units are large, complicated and relatively expensive devices.

Current limiters can limit through-fault contribution by interrupting a fault current before it can reach prospective peak value, but require the replacement of link elements following fault limit operation. They are appreciably cheaper than resonant links.

In the event that none of these options are acceptable, it will be necessary either to reconfigure the system to reduce the fault level at the limiting locations, or to replace the underrated switchgear with equipment suitable for the prospective operating conditions.

An associated problem may arise if the capability of the installed site generator set to withstand the electrical and mechanical effects resulting from a system short circuit is limiting. An appropriately set 'current limiter' can be used to protect the complete generator set itself from damage due to this cause. Alternatively, a mechanical torque limiter can be used to prevent high values of transient torques, produced by the generator, from damaging the shaft couplings, gearbox or prime mover. Both types of device require component replacement following a major fault.

These protective features may also be required when any form of 'auto-reclosing' device is installed in the incoming electricity supply system, whereby the site generator may be left running and generating following interruption of the incoming supply to the site. Under this condition, when the supply is suddenly restored by the auto-reclosing device, the voltages will probably be out of phase with those of the site-generated supply and the generator could suffer a 'double short circuit'.

Alternatively, when the auto-recloser is located near to the generator, the danger can be averted by preventing reclosure while the 'island'-generated voltages are out of phase with the system supply being restored. When the recloser is situated at a higher-voltage section of the system, the out-of-phase voltages will be more significant at the relevant supply transformer through which the site remains connected. Damage can only be prevented by locking out the reclosure when the island-generated voltage at the recloser is out of phase with the incoming supply voltage at the recloser; this can be provided for by a suitable protection relay on the recloser.

Isolation requirements

Regulations relating to the provision of isolation are quite rigid and do present a serious problem in the event of loss of supply from the incoming electricity supply to an island generating system. When the site is capable of operating continuously in island mode, i.e. where the available site generation exceeds the total current site load, or where it exceeds the essential site load when an auto-shed system has reduced the total site load to an acceptable value, then isolation can be difficult. It is not always a simple matter to detect the loss of the incoming supply, or rather the remote disconnection of it. If isolation of the island site does not occur, site generation can feed back into the electrical supply system up to its point of disconnection with its own generation source and can supply any connected load within its total capacity; or it can sustain the system voltage, if there is no connected offsite load, to cause reverse power to flow back into the incomer.

There are many possible permutations of operating conditions that can result from loss of supply to an island site. While in many instances detection of the condition is a relatively simple matter, in some particular conditions detection is very difficult. When the total amount of site generation can never exceed the site load, loss of supply will result in deceleration of the island system. The resulting reduction in generated frequency and/or voltage can be used to operate conventional types of relays to ensure isolation of the site from the incoming supply. However, if the available generation matched an appreciable proportion of the load, the reduction in frequency can be slow and it may be necessary to measure rate of change of frequency. This setting, however, must discriminate against normal healthy transient effects, to avoid unnecessary operation of the isolator. When the site generation equals or exceeds the site load, the normal frequency and voltage will be maintained by the generator set, governor and excitation systems; this will nullify the use of simple relays to detect when isolation should occur. 'Low power' sensing relays may not operate in every instance but, by combining such a device with a similar relay used to detect simultaneous 'low MVAR ', improved detection can be obtained. However, the conditions sensed by these can arise in a healthy system and to prevent such maloperation it is necessary to provide a suitable form of generator excitation control which can ensure that a sufficiently discriminating value of MVAR is always available during healthy conditions.

When a PMS is being used, this can monitor the relevant parameters and ensure that adequate discrimination can be provided by the protection devices and settings included.

Protection discrimination

In normal site load distribution systems it is quite practicable to obtain satisfactory discrimination between the various protective devices installed from the incomer down to the individual loads, thus ensuring minimum disruption in the event of any system fault. However, when a generator is incorporated on the site this no longer applies: a new set of discrimination levels is required to deal adequately with both island mode and parallel modes of operation. When bus section switches are incorporated within the consumer's switchboard to provide isolation for duplicate incomers or duplicate generators, discrimination between these and the other main

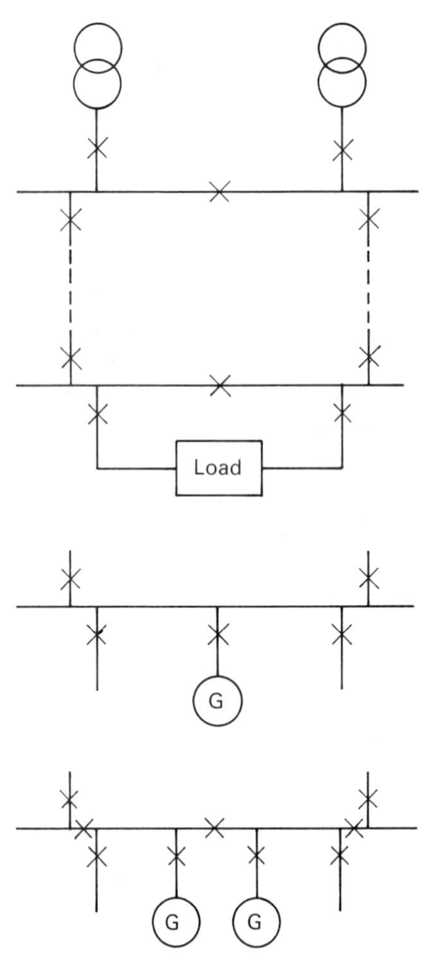

Figure 5.2 *Common duplicate feeder ring systems*

switches becomes more difficult and may require considerable regrading of existing devices (Figure 5.2).

When operation is required with different combinations of incomers and generators, it may be necessary to adjust discrimination settings when changing the configuration. A PMS may be used for this purpose.

Loss of supply

Some important considerations with regard to loss of incoming electricity supply have been discussed above: the effects of autoreclosure following loss of supply; ensuring isolation of the local system in the event of loss of supply; and modifying the protection provided to meet possible configurations arising from loss of supply. There are other factors, however, which must also be considered. These depend upon whether, as a consequence of loss of supply, it is desired to either shut down the site supply or provide a stable island mode operation. The requirements of the former condition are dealt with under isolation of the system, which can be used to initiate a safe shutdown procedure. However, when it is required to continue in island mode operation, while this could be done by shutting down the site as above and then restarting as an island generation system, it is more usual to transfer automatically from operating in the parallel mode to island mode. This involves automatic control activity, requires checking of the transient stability during transfer process, and can involve:

1 **Provision of a stable site supply system** requiring governor and automatic voltage regulator (AVR) control and possibly load shedding and circuit switching. When transfer from parallel to island mode occurs there will usually be a mismatch between the local generated MW and the local MW load. This will result in a site frequency fluctuation, the magnitude of which will depend upon the mismatch of the MW values, the inertia of the island system and the governing capabilities of the prime movers on the site generators and their control logic. Secondary factors will be the relationship between the site absorbed load and the frequency and also the behaviour of the generator excitation system. It is necessary to determine what frequency fluctuation is acceptable to the system while still ensuring that it will return stably to normal steady frequency. A suitable transient stability study can

determine this, together with appropriate governor and excitation control criteria and also what non-essential load must be shed if insufficient local generation is available.

2 **Provision of safety/protection features** to deal with earthing, residual external loads and discrimination. Loss of supply normally also removes the main system earth and it is necessary to close the local generator neutral point to earth through a suitable earthing resistor or other device as quickly as possible. It is also prudent to fit neutral voltage displacement on the generator to ensure isolation if an earth fault should occur while it is still unearthed. It may be necessary to monitor for supply of any system load by the site generator and to disconnect this as quickly as possible by tripping the isolation switch.

Effects of power export/import on supply

Electricity supply distribution systems are normally designed to meet statutory voltage regulation requirements when supplying MW and MVAR and, for economic reasons, they are usually operated close to these limits. The inclusion of a local site generator which can produce a reverse flow of MW or MVAR or both produces changes in the system voltage regulation and this may entail readjustment of existing plant control settings, or even replacement with more flexible equipment to enable the required voltage conditions to be retained.

It is possible to control generator MVAR variations relatively easily, but MW fluctuations are determined by the site load conditions in relation to the prime mover governor control philosophy and cannot usually be controlled arbitrarily, as this would affect site operating economics. If the network is predominantly reactive, the effect of MW load change is not usually serious. In some older networks, in which the resistance effect is considerable, the problem can be more significant and may result in the need to limit MW changes or, alternatively, to change either the electrical system voltage control or its components to meet the new required load conditions.

When operating in parallel with the electricity supply, the local generator should not function using conventional voltage control by means of its own AVR, as this would result in considerable MVAR variations with changes in load. In some instances manual control is satisfactory, but most modern AVRs can be used to control either generator power factor (PF) or MVAR loading, as an alternative to voltage control; this provides a preferable form of excitation control with reduced MVAR variation in the supply. The same device can be switched automatically to function as a conventional AVR when the isolation switch is opened, leaving the generator operating in the island mode.

Metering requirements

Incoming power metering to a consumer is normally unidirectional. When local generation is installed, this will no longer be suitable if any reversal of power is possible. When a new generator is installed there is usually some reconfiguration of the consumer's switchboard, frequently involving the incorporation of a new isolating switch. This can disconnect the incoming electricity supply from the consumer's busbars, which may be required to operate in island mode. This can serve as a useful location for new metering, in some instances. Provision of suitable metering is required to meet the range of possible power import/export conditions of the site, as regards both MW and MVAR. When there are multiple incomers to a single site, these can be dealt with in the normal manner by using appropriate techniques.

Control system for island mode operation

Where a site is never required to operate in an island mode, its control logic is relatively simple. The prime mover governing system is required to ensure the generation of the desired electrical MW output. This will be associated with a corresponding thermal output from the set, together with an effective excitation control system which will regulate the MVAR generated locally and can be used to adjust the imported MVAR, and hence assist in controlling the site voltage.

However, when a site is required to operate in island mode the functions required differ, as the governing and excitation systems are now the sole means of controlling the island system frequency and voltage. The prime mover must provide all the MW required on the island system, and hence any thermal output from the set will vary, depending on the current value of site load. Consequently, other equipment will be required to match the current requirements

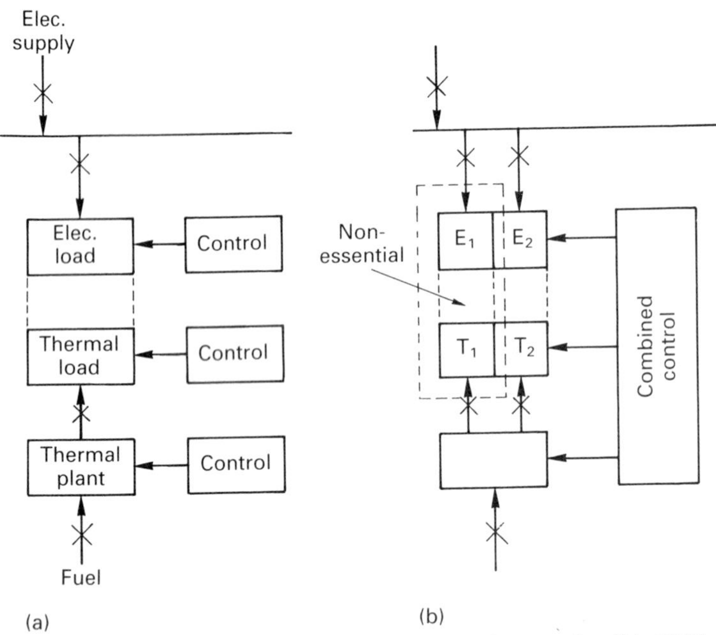

Figure 5.3 *(a) CHP scheme with segregated controls. (b) CHP scheme with integrated controls*

for thermal energy, and the control of this must be integrated with the generator set control and form a single EMS.

In addition, the control system will be required to provide a smooth changeover from island to parallel operation and vice versa, without disturbing the site loading or producing any adverse transient effects on the local system.

Figure 5.3(a) indicates the simplest system, in which separate control of electrical and thermal loads is performed; Figure 5.3(b) represents an integrated control system which combines control of both electrical and thermal loads and, if provided, non-essential load shedding. The more common CHP system, incorporating local generation, is shown in Figure 5.4, which indicates how the generation set waste heat can be incorporated into the system.

It will be clear, therefore, that to provide a satisfactory control system is not a simple matter. It involves many engineering disciplines simultaneously and requires a careful analysis of all the required operating conditions, together with the range of characteristics available from the plant that provides the required site energy.

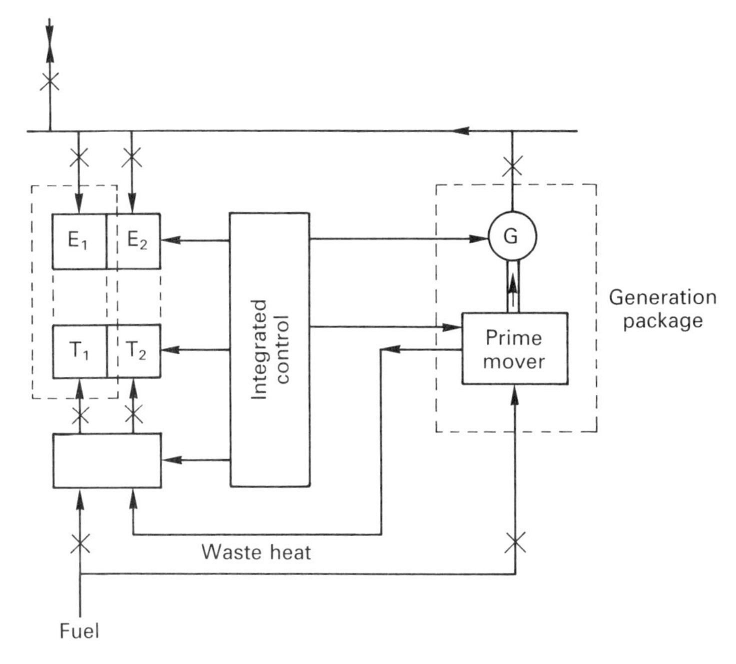

Figure 5.4 *CHP scheme with integrated controls, including waste heat*

When such a control system has been designed and adequately validated in all operational respects – such as being operator friendly, robust, reliable and dependable – it will be necessary to determine the control modes required to obtain optimum operational efficiency. However this is evaluated, it must take into account factors such as economic operation, energy conservation and environmental effects, together with the safety of the plant, personnel and surrounding area.

Special generator sets

As a result of current interest in the principles underlying CHP, including higher energy conversion efficiency, better overall utilization of available energy sources, and reduction of adverse effects on the environment, plant designers have reviewed the experience gained over the years with existing generation equipment. By applying these criteria, they have produced a wide range of machines and equipment which incorporate the reliability of previous designs but are also custom-built to meet the additional requirements of the present day. Every aspect of design of all components has been included in these reviews. As well as employing new and more satisfactory materials and manufacturing processes, use has been made of types of plant and machines once regarded as obsolete but today found to have special advantages for particular applications. This innovation covers the scope of electrical generation sets and includes the use of new types of fuel, special arrangements for combustion in new designs of boilers, improved prime movers of novel design and the incorporation of compound energy cycles which enable higher energy utilization to be obtained. These features have been incorporated gradually over the years as such new designs have become economically viable and proper commercial assessment has justified their installation. The

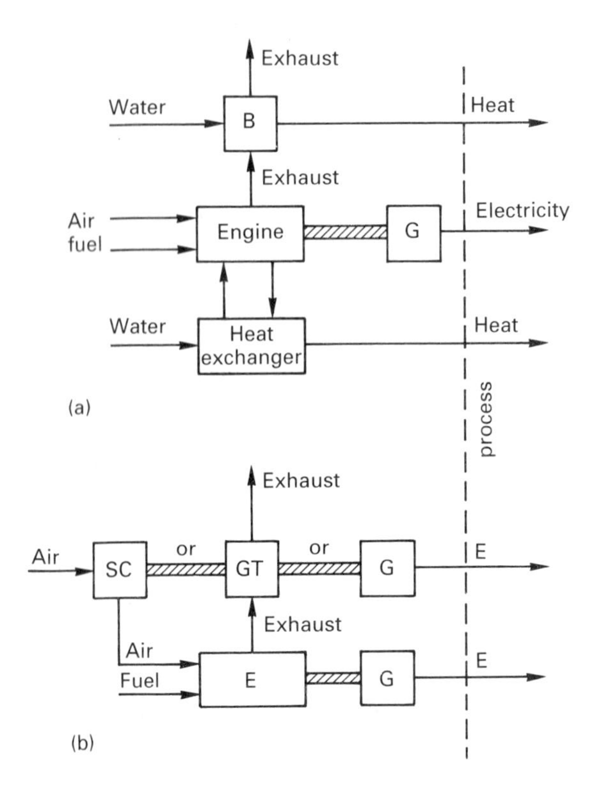

Figure 5.5 *(a) Simple flow diagram for engine-driven generator. (b) Simple flow diagram for engine-driven generator incorporating an exhaust turbine to drive a supercharger or generator*

process has been assisted in the UK by government department initiatives.

Some of the simple improvements that can be made to a basic generator set are indicated in Figure 5.5. Many permutations of these can be made by selecting the particular features that can result in plant optimally designed for a specified set of operating conditions and meeting strict performance guarantees.

Figure 5.5(a) indicates how the high-grade heat energy in the engine exhaust can be used to raise steam, or heat water in a suitable boiler (B), while the low-grade heat can be removed by cooling the engine oil by using it to heat circulated water (CW). It is also possible to absorb mechanical energy from the exhaust by using it to drive a turbine which can be coupled to a supercharger (SC) to raise the inlet air pressure and increase the flow supplied to the engine; alternatively, it can be used to drive an electrical generator, as in Figure 5.5(b). With this system it is also practicable to recover heat from the exhaust.

The two simple types of steam turbine (ST) are shown in Figure 5.6(a) for condensing and Figure 5.6(b) for back-pressure sets. If it is required to supply boiler steam for process purposes, it is advantageous to match the pressure supplied to the turbine. The turbine exhaust steam is passed to a condenser (Cd) to give optimum pressure drop in the turbine, and the resultant condensate can be returned to the boiler as feed water. The medium used to remove heat from the condenser is normally water, but the same process can be performed by cooling towers. In both instances the heated medium can have adverse effects on the environment (such as raised river water temperature on discharge), but it is generally uneconomic to recover this low-grade waste heat.

However, by using a back-pressure set, the boiler can be used to supply high-pressure steam (HP) to the turbine which exhausts at a pressure matched to the requirements of the process with regard to pressure and mass flow. This enables more efficient boilers and turbines to be utilized, and provides a source of high-pressure steam for process purposes when this is also required. Where several pressures are required by the process, multi-stage turbines are available from which steam at a suitable pressure can be bled, giving a greater flexibility in application as shown in Figure 5.6(c).

There are similar situations in possible gas turbine (GT) cycles. The simple cycle shown in Figure 5.7(a) can be used to give an

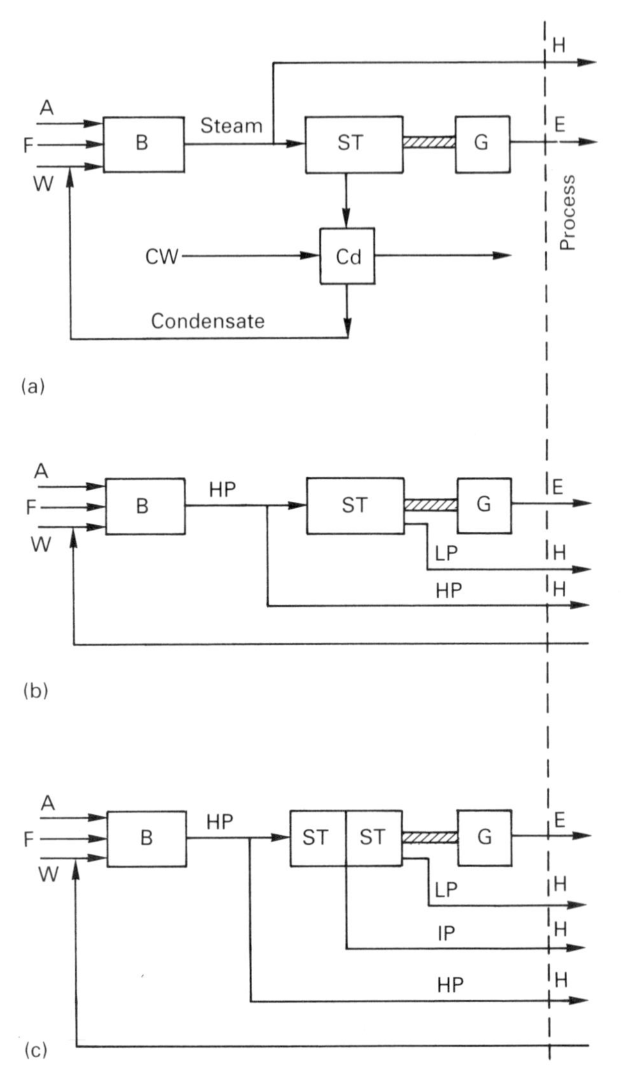

Figure 5.6 *Simple flow diagrams for steam-turbine-driven generators. (a) Condensing set. (b) Back-pressure set. (c) Multi-stage turbine set*

efficiency in the region of 35–40 per cent, depending on the turbine type and design. The performance is mainly affected by the compressor (C) pressure ratio, which determines the air mass flow, and the working temperature of the gas. Improvements can be made by using a regenerative cycle (Figure 5.7(b)), in which the turbine exhaust is used to preheat the compressed air prior to

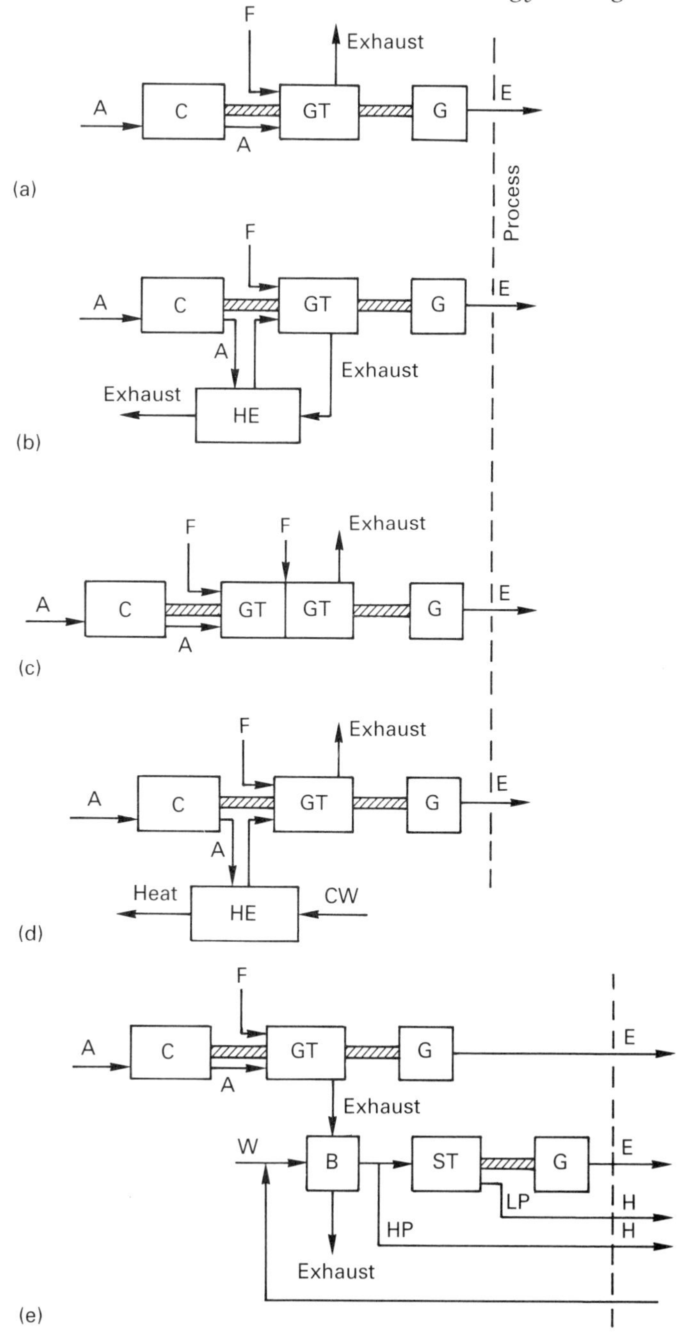

Figure 5.7 *Simple flow diagrams for gas-turbine-driven generators. (a) Simple cycle. (b) Regenerative cycle. (c) Reheat cycle. (d) Intercooler cycle. (e) Combined generation cycle including steam turbine generator set*

combustion, hence reducing the amount of fuel required to reach the same working temperature. With high compression ratios, however, the air temperature is raised to such a value that it reduces the mass flow; the use of an intercooler to remove heat from the compressed air and thereby increase its mass flow can raise the overall turbine efficiency (Figure 5.7(d)). When very high ratio compressors are used, sufficient oxygen may remain in the gas mixture after passing through first-stage combustion to support the combustion of additional fuel. Figure 5.7(c) indicates such a reheat cycle.

Obviously, suitable combinations of these features can be made, depending on the machine design and required performance. In all cases, simple exhaust heat recovery can be provided from a boiler fired by the exhaust gases, and the availability of such steam can be used to advantage if steam injection is used in the gas turbine at the combustion stage to reduce the amount of NO_x produced in the exhaust as well as increasing the output from the turbine. A logical development from the above is a combined cycle (Figure 5.7(e)), in which the gas turbine exhaust fires a boiler which, in turn, supplies a steam turbine, both turbines driving electrical generators. The other cycle improvements can be added to further improve the performance of such cogeneration cycles and efficiencies above 70 per cent can be obtained.

Specific site requirements can be matched by a judicious selection of plant and energy cycles; features such as secondary firing of boilers normally heated by turbine exhaust gases can assist in energy balance control and also provide an emergency or back-up feature in the event of loss of one prime mover.

More exotic cycles have been developed for special applications; for example, modern ship propulsion schemes tend to be integrated with the vessel's total energy demand, including electrical and thermal. Such an installation could consist of a slow-speed diesel engine directly driving the propeller but also having a shaft-mounted generator which could be used to produce electrical energy when required, but could also act as a low-power propulsion motor in an emergency. The engine exhaust gases can be used first to drive turbines coupled to an electrical generator and a supercharger (SC), and then to raise steam in a suitable boiler, provided with secondary firing, which can drive a steam turbine coupled to an electrical generator. Auxiliary diesel-driven electrical generators would be required to provide power for system start-up and also to meet emergency conditions (Figure 5.8(a)). Such a system could be used to supply the range of propulsion powers required, which vary with the vessel speed and hence propeller speed, together with all the vessel auxiliary services supplied from the electrical system. These would include the energy system auxiliaries, the domestic

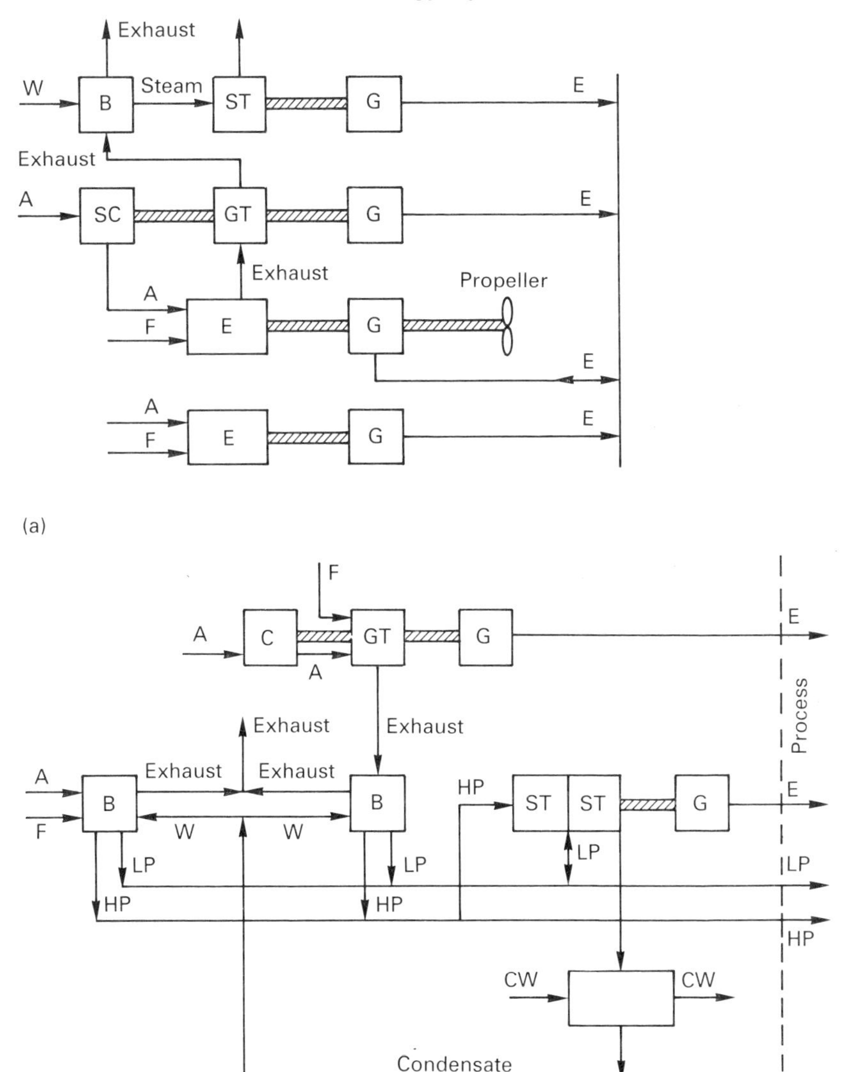

Figure 5.8 *Simple flow diagrams for composite systems. (a) Possible ship propulsion system. (b) Possible process plant requiring large variations in steam demand*

load and special vessel loads, which could include a significant refrigeration load, which would require energy generation even when no propulsion power was being used.

Other special requirements could be quite different, e.g. a process plant requiring a continuous supply of energy comprising a large variable demand for steam at both high and low pressure (LP) as well as a considerable electrical load. Figure 5.8(b) indicates one possible combination to meet such conditions. It incorporates the sets shown in Figure 5.6(c) and Figure 5.7(e), together with a separately fired boiler providing an alternative source of low-pressure steam which can be used to supply the condensing steam turbine with a pass in/pass out facility. This gives greater flexibility to meet variations in process steam demand.

Another application for cogeneration of increasing interest is the straight combined cycle generation plant of which high efficiency and low power output cost are the prime targets, with only a secondary purpose of supplying steam to prospective customers. Such sets, being designed for continuous high-power output, can produce very high efficiencies, as the range of operating conditions of each item of plant is small and optimum design and operation are possible.

All these special cycle systems require careful analysis of operating conditions, including their range of variation, in order to determine the necessary design characteristics of each separate item of plant to ensure full compatibility and also to meet the final requirements of high efficiency and low environmental contamination. This operation involves several disciplines of engineering. While single items of plant can be designed by specialist engineers, the coordination of the complete system requirements requires the services of application engineers with wider experience. In general, a design teamwork approach is required to ensure full compatibility not only in performance ranges but also in the fields of reliability and safety.

Increasing use is now being made of simulation procedures to facilitate the study of integrated systems. This involves models of the process itself as far as it affects the energy requirements, together with electrical system modelling referred to in Appendix 3.

When a suitable plant has been designed, proved and installed, simultaneous control of all the variable parameters is required to ensure continuous stable operation, at optimum performance, for each separate operating condition imposed on the individual items of plant within their operational capability; this requires the inclusion of a suitable EMS. Where a full simulation of the appropriate system has been carried out, the necessary control logic may have been determined and proved and the EMS logic obtained. Validation of the simulation used for the various items of plant and control will then ensure satisfactory operation of the plant.

As discussed previously, it is also necessary to incorporate into the EMS not only the necessary steady-state logic required for system control, but also that required to ensure that, in any emergency, it will perform the essential control actions to restore the system to a safe, stable state, even if this should require shutdown of some or all of the plant.

Another, equally important, function of the EMS is to control the plant so that it operates at optimum economical values of parameters for the particular loading conditions, on the basis of fuel/energy costs and taking into account factors such as first or capital costs, depreciation, reliability of plant and the consequences of failing to meet operational demands. These often prove to be the most difficult features on which to obtain firm and accurate data.

Cost factors

A knowledge of cost factors is required, initially to enable a suitable system to be selected and designed together with an appropriate control system. This requires prediction, not only of the behaviour of the system itself, but also of the loadings to be supplied in the future, and finally establishment of the variation in costs of energy over the same time interval.

For any particular system there will usually be numerous alternative possible configurations of plant and modes of operation which will meet the specified requirements, involving different types of fuel, prime movers, generators, boilers etc. together with necessary auxiliaries and also involving various numbers and sizes of such units, either duplicate or of differing capabilities where greater flexibility is required. At this stage there are some factors which will be overriding and enable a significant reduction to be made in the number of practicable alternatives before attempting to carry out a detailed costing analysis of the alternatives. Thus a suitable supply of locally available cheap fuel may justify the use of gas turbines on offshore oil/gas platforms.

One of the main factors in all installations is the cost of providing adequate alternative energy sources to the site; this is related to magnitude of demand, which can significantly affect cost per unit at specific quantities and which might justify mixed energy source supplies, giving the added advantage of increased reliability. The actual site energy demands in all forms must then be related to the possible sources and to the plant being considered.

Actual site load requirements as a function of magnitude and time must be converted to suitable power/energy forms to enable a realistic assessment to be made for costing purposes. Depending upon the nature of the load, this can be done by one of the conventional procedures already discussed under generator rating. Most process loads tend to follow time cycles, the 24-hour cycle being the most common; however, the effects of weekends and summer or Christmas holidays are often equally significant. It is

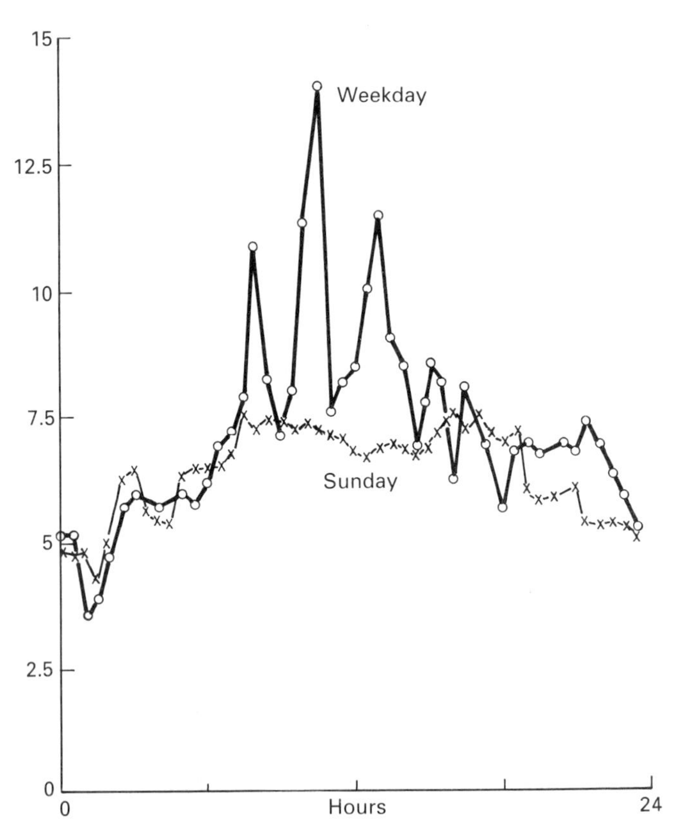

Figure 5.9 *Typical daily variation of steam demand*

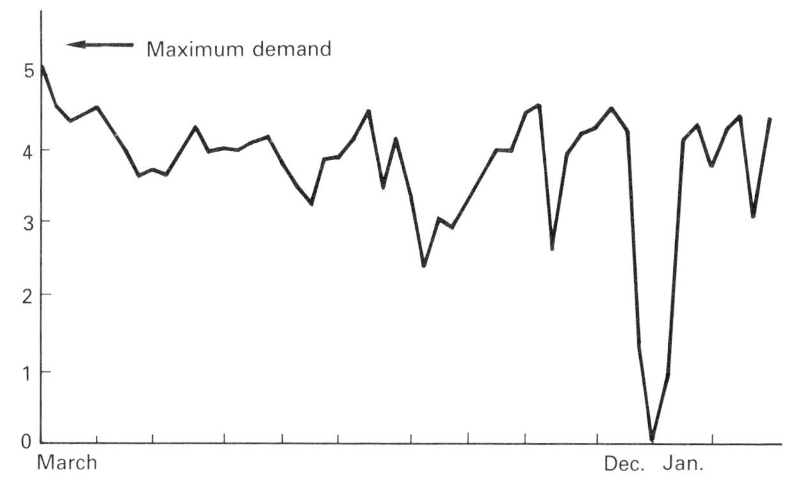

Figure 5.10 *Typical annual variation of steam demand for UK process plant*

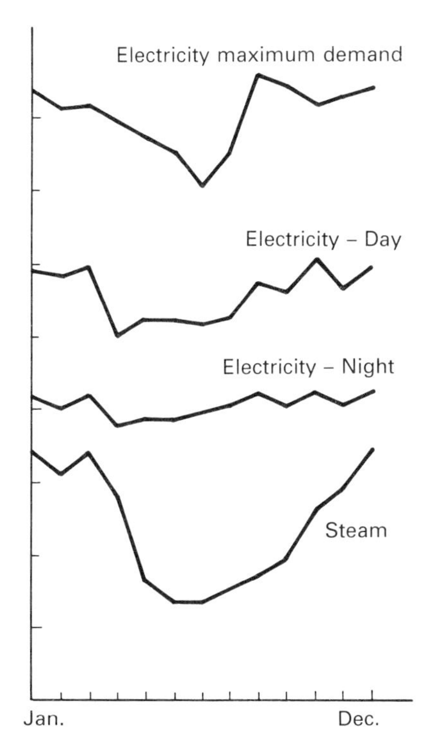

Figure 5.11 *Typical annual demand for electrical power and steam for UK domestic-type plant*

important to establish such effects, as they may correspond to special energy and operating charges applicable at such times (Figures 5.9 to 5.12). The magnitudes of load fluctuations are also very important, as the plant will be required to operate over the complete range. This factor may affect the type of plant required, with regard to characteristics and mode of control. Thus it may be economical to amend the plant energy cycle in such a manner as to provide for peak demands by using a modified control sequence resulting in reduced efficiency for the relatively short times involved, but offsetting this possible increased cost by using equipment having an appreciably lower capital cost than if it had been designed to meet the maximum demands continuously at optimum efficiency.

As well as considering load fluctuations it is necessary to consider the effects of future variations in fuel and energy charges. This is a much more intractable problem, but is particularly important when comparisons are being made between projected installations which

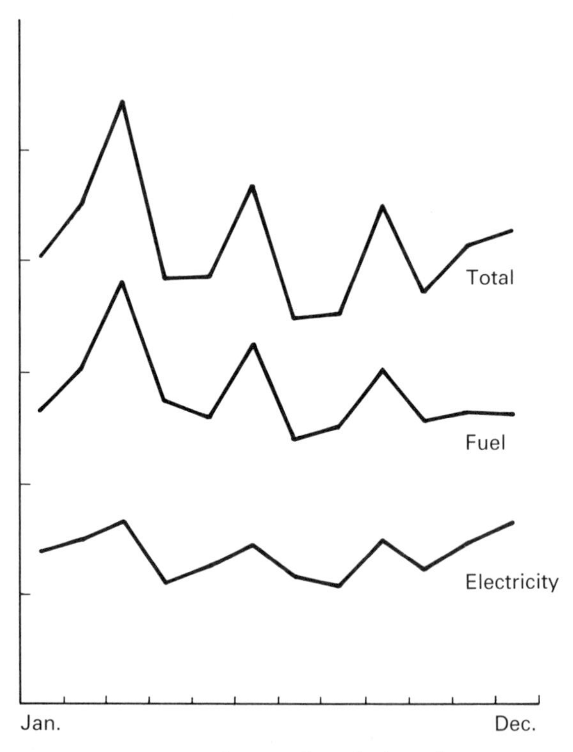

Figure 5.12 *Typical annual variation of energy costs in the UK*

could use different forms of energy, or mixtures of fuel or energy in different proportions. Some measure of estimation of future costs can be obtained from energy suppliers. In the present situation in the UK, where alternative fuels are under the control of competitive private enterprises, it is not possible to rely on long-term comparative forecasts of the costs involved (Figure 5.13). Until a national policy on energy is available which can establish such relative costing between alternative fuels, it will be necessary to continue to include a variation factor on predicted costs to cover a reasonably practical possible variation on the firm projected costs. This particular factor can prove disadvantageous to plant cycles dependent for their good performance on one specific form of energy, such as a steam-turbine-driven generator supplied with steam from a coal-fired boiler which can only operate when an adequate supply of suitable coal is available. This often results in preference being given to combined-cycle generation, which can be operated efficiently using alternative forms of fuel or energy. Alternatively, a gas-fired boiler supplied from an uninterruptible gas supply could be arranged for emergency oil-firing in event of non-availability of gas.

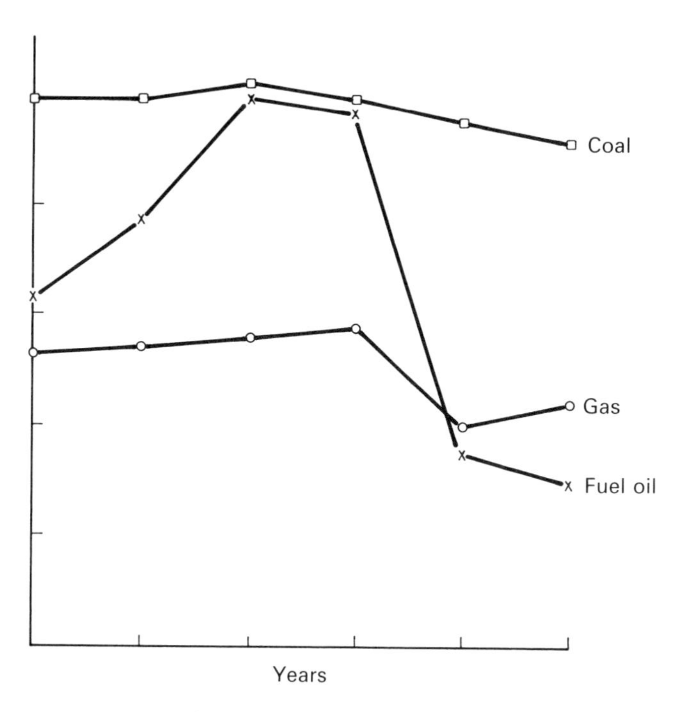

Figure 5.13 *Relative fuel costs over a 5-year period in the UK*

This, although more expensive in fuel costs, could ensure continuity of electrical generation which might prove more financially beneficial than shutting the generator set down with loss of power to the system, and possible shutdown of process plant and loss of production.

Some measure of control of future fuel costs can be obtained by arranging long-term contracts with energy suppliers who can establish future variations. Gas contract schedules can be obtained for medium terms of 3–10 years, and even longer-term interruptible schedules are available to meet the requirements of the large generation plants.

After a plant has been installed, proved and put into service, it is possible that there will be differences between the predicted behaviour and that actually experienced. It will be necessary to evaluate the significance of these with regard to the technical performance of the plant and system and their economic performance: it may be advantageous to review the control logic adopted in order to improve the overall performance.

When a system is operating satisfactorily it will be necessary to monitor the actual costs encountered and to compare these with the predicted values. If there is a significant deviation, but one which is within the projected range of values in the sensitivity analysis, then any corrective action envisaged can be implemented in the mode of operation and control. However, should the cost deviations exceed those originally projected, a full reassessment of the system should be made to ensure that satisfactory and economic operation can be restored, and that it can also be retained during foreseeable cost excursions in the future. The nett operating costs of a system can be determined from actual costs over any period but this may not determine its overall economic viability, since future expenditure could reverse profitability. It is necessary, therefore, to ensure that all chargeable costs are within the appropriate projected range, including all contingency amounts. Thus, while it is important to include accurate costs in any calculations on which time-dependent control functions depend, it is the integrated values that are relevant to overall financial performance; consequently, the procedures for evaluating prospective costs differ from those applicable to control logic.

Cost investigation procedures can vary from very simple to very complex. It is usual to initiate feasibility studies on a simple cost basis as a starting point for any project. Thus the addition of a gas turbine generator set and waste heat boiler to an existing structure, to provide the basic site electrical load plus process steam requirements, can be assessed as a capital cost of installation with three main component costs:

Main plant, generator set, boiler etc.	65 per cent
Civil work, piping, electrical, installation	25 per cent
Design, engineering etc.	10 per cent

Comparing the pre-CHP and post-CHP annual energy costs as percentages of the total capital cost:

	Pre-CHP (%)	Post-CHP (%)
Electricity	50	1
Gas (boiler)	34	–
Gas (gas turbine)	–	58
Fuel oil	1	–
Distillate oil	–	1
(required for gas turbine during gas interruption)		
Total	85	60

This indicates an annual saving of 25 per cent of the capital cost, which gives a four year payback period. Figure 5.14 presents in simplified form a sensitivity analysis of the principal fuels, allowing cost variation ranges of +50 per cent to –20 per cent on gas and +20 per cent to –10 per cent on electricity. From this it can be seen that within these ranges the extreme variation of savings is between 40 and 160 per cent of the nominal saving. Such diagrams can be produced with finer scales to enable corrections to be made to estimated savings for any required fuel cost variation.

When a satisfactory feasibility study has been carried out, it is then necessary to refine the cost estimate and also to consider alternative plant arrangements which could offer operating advantages. Such comparative analyses should include all known factors to a reasonable degree of accuracy and must be based on identical loading conditions. Several extreme possible loading conditions, such as where the power-to-heat ratio is significantly different,

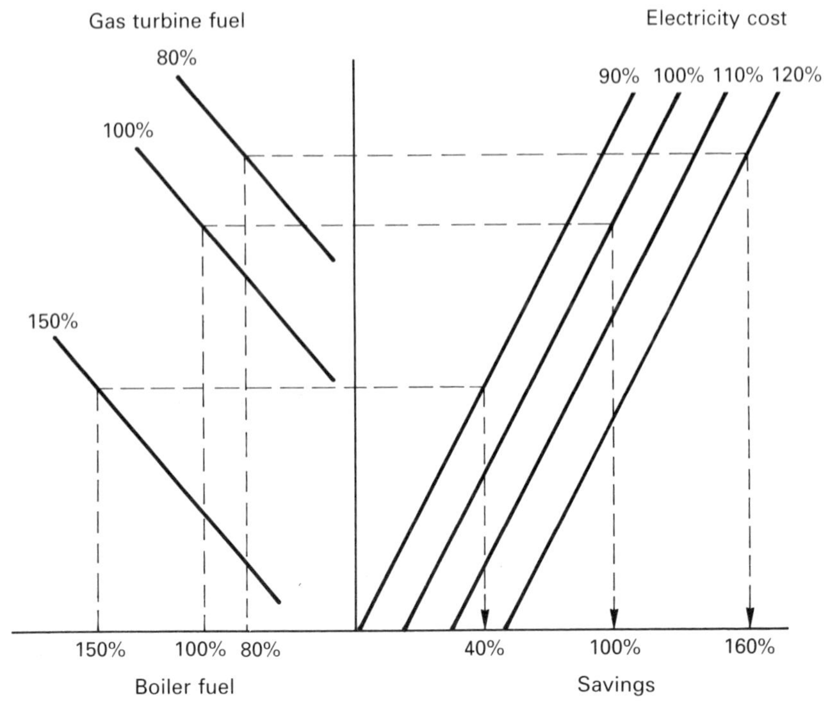

Figure 5.14 *Sensitivity analysis for fuel cost variation*

should be assessed as they could affect the performance of alternative schemes very differently.

At present, costing models or programs are available which can evaluate the operating costs and associated functions of systems consisting of specified items of plant for defined loading conditions which have been used by the plant designers to determine the required performance to ensure correct system behaviour. Such programs can be used to evaluate different plant combinations for the same plant loading conditions; they can also be used to quantify the consequences of varying the characteristics of individual items of plant within a system, which can lead to the most economic selections of components and their operating characteristics for any system. They can further be used to determine the effects of variation of system loading on a particular configuration of plant, and hence to determine the most economic means of using the plant selected.

When designing a CHP scheme it is customary to prepare energy diagrams to identify all the details of distribution of electrical loads as well as heat flows, which also indicate variation in these parameters. In any process plant there exist functional relationships between the various energy-consuming units and these must be matched by the energy-producing sources; this involves the energy monitoring and control systems.

These must maintain the desired demand while operating the generating plant within its designed operating range. In all but the simplest of installations, this operation requires automatic features which may be applied to individual plant units or, if it is desired to optimize the operating efficiency of the plant overall, require some form of integrated control. It is a natural evolution to add cost to the functions involved in the control system and to use plant economy as a major control criterion. Such a control system requires a databank including, at least, all the items of cost used for the projected system evaluation, updated as far as possible to represent actual current values.

By combining the functions available from the process control system model and the costing system model into a single control, it is possible to produce a fully automated system organized to give the most economic consumption of energy for the required plant demand. The effectiveness of such a system depends upon the accuracy of the data available to it. All data must be kept up to date or at least kept within the working tolerances used by the control system, e.g. the range of acceptable calorific values of the fuel in use.

With such sophisticated systems it is practicable to incorporate any relevant operational safety limits, and also any limits such as those applied to emission products to limit environmental pollution to acceptable values. The control system can use such limits to override other normal limits, including cost control criteria, if necessary, and advise the operator that such action has been initiated.

Total energy

It is now usual, when investigating a new CHP project, to undertake some form of total energy analysis, as this could identify factors that significantly affect the choice of type of CHP scheme. The first consideration in any total energy study is to try to determine energy loss or wastage, in any form, and to establish how this can be eliminated or reduced within the existing system. Such features could include the shutdown of non-essential plant which may be left running to avoid the trouble and inconvenience of shutting down and restarting, and basic operations such as improving thermal insulation of piping etc. and the reduction of leakage from steam, water or fuel lines.

When the limit of energy efficiency improvement is reached in this manner, the use of alternative energy sources should be considered, such as solar radiation. However, the most useful source is the recovery of waste energy from the process. This usually appears as low-grade heat; devices such as heat pumps can be utilized to make some of such heat usefully recoverable. Other devices are available which can perform similar functions and, although in the past they have been disregarded as being an unnecessary complication involving capital expenditure and a requirement for maintenance, the new attitude to energy conservation has led to a reconsideration of such devices and procedures. Energy engineering is fast becoming a useful and effective engineering discipline.

The concept of total energy can be seen implemented in the recent design proposals for large combined-cycle generator sets and generating stations, where the principles can be applied at the conceptual stages of design, meeting demands for the economic utilization of available energy sources, and reducing waste and adverse effects on the environment.

Appendix 1
Induction generation versus synchronous generation

Electrical AC generators

Rotating electrical generators involve the conversion of mechanical energy to electrical energy and this is performed in the air-gap space between the stator bore and the rotor periphery when the rotor is driven by some prime mover capable of providing mechanical power. When rotating at speed the rotor possesses stored kinetic energy by virtue of its inertia, but this energy cannot be transferred directly to the stator. However, if a suitable magnetic field, commonly known as magnetic flux, can be produced in this air-gap, it is possible to induce a voltage, by electromagnetic induction, in conductors on the stator or rotor if they are moving relative to this field. Inverted machines, in which the outer unit (stator) is driven while the inner unit remains stationary, are available; these function in the same manner electromagnetically. Further consideration will be confined to the conventional arrangement which forms the vast majority of AC generators.

A suitable magnetic field can be produced in the air-gap by passing current through windings on the rotor, or stator, provided that they are correctly designed for the required conditions and electrical power supply available. Permanent magnets can be used for this purpose but, at present, they are only suitable for low output machines and will not be considered further.

When a rotor is driven at 3000 rev/min and is provided with one pair of poles, and its winding is supplied with DC, it will produce a magnetic field rotating at the same speed as the rotor. This produces a field alternating at 50 Hz when viewed from the stator.

However, if the rotor winding had four poles and operated at 1500 rev/min it would still produce 50 Hz on the stator, and by suitable selection of rotor winding pole number it is possible to obtain suitable stator frequency for a large number of rotor precise speeds. Such DC windings can be mounted on salient poles on the rotor, or can be wound in slots on the rotor periphery in a concentric distributed winding arrangement. Synchronous machines with these types of windings are usually described as 'salient pole' or 'round rotor' machines.

This winding, known as the field or excitation winding, requires energy to produce the field and the current is fed into it from an electrical power source, usually outside the generator, but commonly from an exciter connected to the generator shaft, from which mechanical energy is converted to electrical energy in a similar manner as with the main generator. Variation of the magnitude of the excitation current produces a corresponding change in the voltage induced in the stator winding.

It is also possible to produce a rotating magnetic field by passing AC currents, usually polyphase, through a suitably distributed winding on the stator. Such a winding can be arranged to produce one or more 'pole pairs' and, if the current is provided from a 50-Hz power source, it can produce a magnetic field rotating at 3000 rev/min, 1500 rev/min etc. in the same way as the rotor winding using DC current. The production of this field also requires the provision of reactive power, or kVAR from the source of AC current.

Synchronous generators

By providing an electrical machine with a rotor winding producing a rotating magnetic field corresponding to 50 Hz when driven by the prime mover at rated speed, and also a stator winding with the corresponding number of poles, this latter will produce a 50-Hz voltage. However, if a 50-Hz current is allowed to circulate in this winding it will produce a second magnetic field rotating at the same speed as the rotor-induced field: the interaction of these two fields can result in the transfer of mechanical energy from the rotor to electrical energy on the stator. Since these two fields rotate in synchronism, such a machine is known as a synchronous generator.

If a synchronous generator has a resistive load connected to it when it is operating at rated speed and generating a voltage, current will flow in the stator windings in phase with the voltage. The second electromagnetic field produced in the air-gap will not be in phase with that produced by the excitation winding, although it is synchronous with it, and the interaction of these fields will transmit energy across the air-gap in the form of real power (kW) equal to the power absorbed by the resistive load. Initially the power required is obtained from the kinetic energy of the generating set and causes a transient deceleration which will result in the speed-governing system increasing the energy input to the prime mover. This will cause it to reaccelerate and restore the set to its steady operating condition at synchronous speed, but with an appropriate angular displacement (load angle) between the two component air-gap fields.

The application of load to the generator will also cause an instantaneous voltage regulation drop due to the load current flowing in the generator impedance; if an automatic voltage regulator (AVR) is provided, this will adjust the excitation current and restore normal operating voltage. The generator will then be operating again at rated frequency and voltage and the current will be in phase with the voltage, i.e. at unity power factor, providing the load kW. The displacement between the rotor-induced and stator-induced fields is known as the load angle.

When the generator load current is purely reactive, the electromotive field produced by this is in phase with that produced by the rotor excitation, i.e. at zero load angle, and the relative polarity of the two fields will depend upon whether the generator current is reactive lagging or leading the voltage. When the generator power factor is lagging, the two fields are in opposition and if leading they will be cumulative. At the instant of applying such a load the net electromagnetic field will result in a generator voltage which is respectively reduced or increased and the AVR will then adjust excitation current, to restore normal voltage.

Normally, loads consist of an impedance comprising both resistive and reactive components and both the above effects occur simultaneously. The simple vector diagram (Figure A1.1) indicates

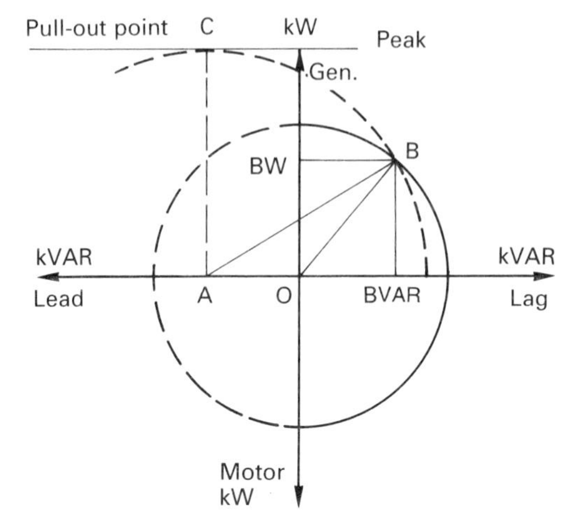

Figure A1.1 *Simple synchronous machine vector diagram kW–kVAR*

the typical resultant steady states for variation of kW and kVAR, where the vertical axis indicates generated kW and the horizontal axis represents reactive kVAR, leading to the left and lagging to the right. It is also possible for a synchronous generator to act as a motor, which would be indicated by operation down the vertical axis. Operation at zero load, i.e. at point O, corresponds to a value of excitation to produce rated voltage at rated frequency and this is represented by AO. When operating at rated load (point B, with OBW power and OBVAR reactive lagging), the corresponding excitation is represented by AB and the value OB represents the amount of excitation that would be required on the rotor to equate with the field produced by the current in the stator winding.

Operation at constant excitation is indicated by the broken-line circle, centre A, radius AB, and operation is possible in this whole region. The value AC indicates the greatest amount of kW power that can be transmitted across the air-gap for a value of excitation corresponding to AB. It should be noted that this simplified operation diagram only applies to steady-state operation at rated frequency and voltage. The radius OB represents rated kVA in the generator and hence the operating locus of the circle produced by this represents the steady-state thermal limit of the machine. It will be appreciated, therefore, that operation at kW greater than the rated value requires reduction in kVAR and vice versa for constant excitation. When AB represents the maximum loading that can be applied to the generator to remain within the steady-state thermal limit, then this diagram gives the limiting steady-state operating loci and is called the 'capability diagram'. In practice, when an AVR is used, the excitation will be adjusted to match the diagram corresponding to the actual load kW and kVAR.

Operation under other conditions, such as values of frequency or voltage other than rated values, can easily be deduced from this simple diagram.

Induction generators

If the polyphase stator winding is connected to an electrical power source of suitable voltage and frequency it can produce a synchronously rotating magnetic field in the air-gap of a machine. Rotating the rotor at the same speed will produce no energy transfer; however, if a suitable distributed polyphase winding were provided on the rotor and the rotor speed differed from that of the rotating field, then this field would produce a voltage in the rotor winding. The frequency of this voltage would equal that of the stator electrical supply only if the rotor were stationary, the machine acting in a similar manner to a transformer or induction regulator: as the difference in speed between the rotor and the air-gap rotating field reduced, the rotor-induced frequency would be reduced in proportion with the difference or slip. The term slip is used to express the fraction of rotor synchronous speed which corresponds to the difference between rotor synchronous speed and rotor actual speed. Slip is normally regarded as positive when the rotor is running below synchronous speed and negative when above it.

When the rotor operates at slip speed a voltage is induced in it at a frequency equal to the product of slip and synchronous frequency. If the winding is closed, either by being short-circuited as in the familiar 'cage' induction machine, or through some impedance, such as can be used with a 'slip ring' type of machine, then current will flow in this winding and this will produce a second electromagnetic field in the air-gap. As in the case of the synchronous machine, the interaction of these two fields can transfer energy across the air-gap between the stator and the rotor, and the direction of this flow depends upon whether the slip is positive or negative. Positive slip can transfer energy from the stator to the rotor whereas negative slip can transfer energy from the rotor

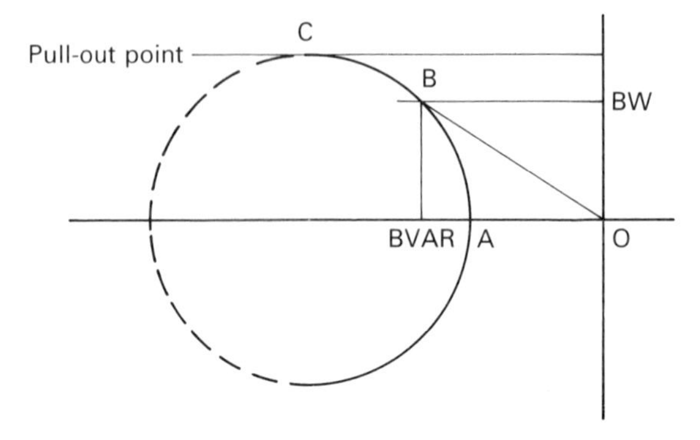

Figure A1.2 *Simple induction machine vector diagram kW–kVAR*

to the stator, i.e. the machine changes from a motoring mode to a generating mode.

Increase in slip magnitude increases the induced voltage and current in the rotor windings and hence the strength of field produced and also the magnitude of the transmitted energy. A typical characteristic is shown in Figure A1.2, where OA represents the reactive power drawn from the supply to produce the basic air-gap magnetic field, or flux; this corresponds with the value OA in Figure A1.1 which is produced by the excitation on the synchronous machine rotor. OB represents the rated operating condition for negative slip, i.e. with the machine operating as an induction generator. Further increase of slip is represented by movement round the operating locus to point C, which represents the greatest amount of kW that can be transmitted across the air-gap.

Comparison of characteristics

A comparison of Figures A1.1 and A1.2 clearly shows one of the major differences between synchronous and induction generators. Induction machines always require reactive power (kVAR) from the power supply, whereas the synchronous machine can either provide reactive power to the supply, which is its normal mode of operation, or can operate with zero kVAR flow – or could even, if desired, operate by drawing reactive power from the supply.

The other major difference between the two types of machine, however, is in the electromechanical requirements: the power transmitted by the synchronous machine is dependent on the load angle, which is the phase angle difference between the synchronous field in the gap and the rotor, which continues to operate at synchronous speed, whereas the induction machine has to have a 'slip', i.e. a difference in speed between the synchronous field in the gap and the rotor, and the magnitude of this slip must change to change the power flow.

Other significant differences between the two types of generators are associated with their mechanical differences and with operational procedures.

Air-gap

The radial length of air-gap, i.e. the distance between the rotor periphery and stator bore, is an important design factor both mechanically and electrically. The amount of kVAR required to produce the air-gap electromagnetic field is roughly proportional to the length of this air-gap. Figures A1.3 and A1.4 illustrate the approximate effects of halving and doubling air-gap length on the original characteristics (Figures A1.1 and A1.2), assuming that other factors remain constant.

With the induction machine the effect is merely to reduce kVAR or increase it, which results in a slight improvement or impairment of the operating power factor. In practice the air-gap is made as

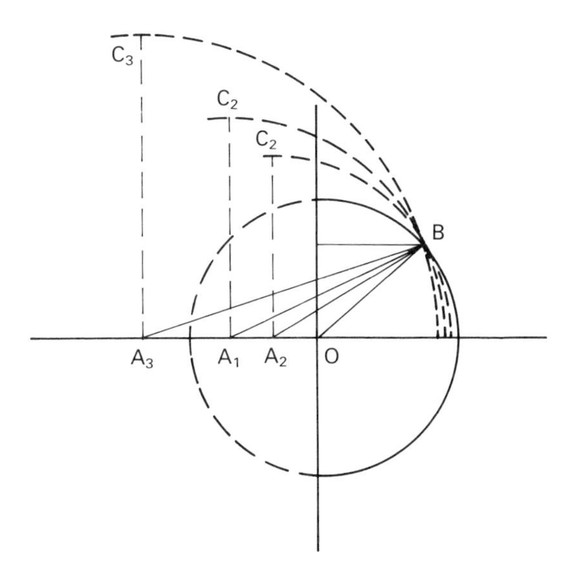

Figure A1.3 *Effect of air-gap on synchronous machine characteristics*

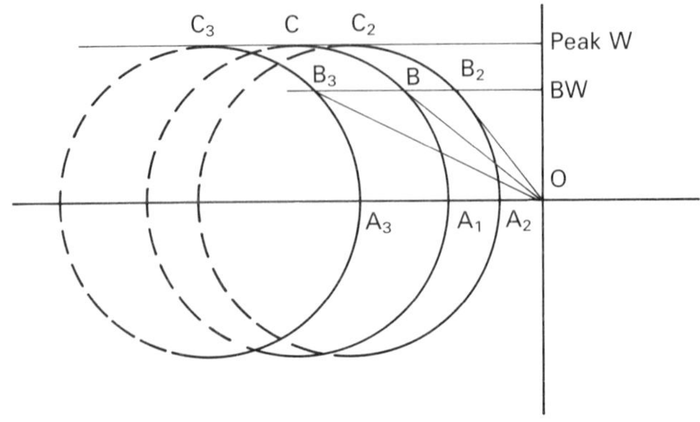

Figure A1.4 *Effect of air-gap on induction machine characteristics*

small as possible consistent with safe electromechanical operation and reasonable losses, to give the best power factor. The smaller the air-gap, the greater the friction and windage loss and also the parasitic magnetic and electric stray load losses which reduce efficiency. The stray magnetic field in the air-gap, in addition to transmitting power across the gap, exerts mechanical forces on the rotor but when the stator and rotor are truly concentric these balance out and normally have no effect. However, if the air-gap is not uniform round the periphery the radial force will also vary, and if there is any appreciable eccentricity there can be a significant radial unbalanced magnetic pull between the stator and the rotor. The magnitude of this is dependent on the amount of air-gap eccentricity and therefore, for a given rotor displacement, the unbalanced pull is greater when the average radial air-gap is smaller. The importance of this factor is affected by the construction of the induction machine, particularly the type of bearings used; the stiffness of the rotor shaft can be very significant for some induction generators.

With the synchronous machine, however, variation of the air-gap merely changes the amount of excitation required for the machine and does not affect the operating power factor. It does affect the amount of excitation power, which affects the efficiency. In addition, it affects the maximum kW power that the machine can provide or absorb under steady-state conditions. When this value is exceeded, the synchronous 'load angle' increases until the machine becomes unstable.

A synchronous machine is usually provided with an automatic excitation control system which enables it to give a peak power capability, known as the 'pull-out torque', or power, corresponding to its maximum available excitation power. This enables the machine to operate as a generator at constant voltage over a wide range of loads and power factors, or as a motor at constant power factor or constant kVAR, or to meet any desired relationship of kW and kVAR over the range of operation within its available excitation power locus.

The choice of air-gap length for a synchronous machine is not made solely for electromechanical reasons and is usually selected to give a desired value of peak power capability, and, hence, machine impedance. Thus the air-gap is usually found to be very much larger than that of the equivalent induction machine and the effects of unbalanced magnetic pull and stray losses are very much reduced.

Costs

Even when the synchronous machine excitation loss is taken into account, it is found that its efficiency can be greater than that of the corresponding induction machine. This factor, together with the available option of selecting the operating power factor (PF) of the synchronous machine, can render it more economical to operate under specific conditions. The first cost of a synchronous machine is usually greater than that of an equivalent induction machine because of the need to provide the special excitation winding and an exciter or other source of excitation power.

Any accurate comparison of costs of induction versus synchronous machines must include the first cost plus the operating costs, based on the actual load conditions and control procedures to be adopted. It must also allow for the costs associated with factors such as possible benefit of controlling operating PF or kVAR and structural costs incurred in meeting the levels of accuracy required during erection and line-out requirements and in avoidance of maloperation due to foundation movement, seismic or other shock forces, or structural movements such as occur on board ship or on offshore platforms.

Dynamics

Another significant difference between the two types of machine, and one which cannot easily be assessed financially, lies in their dynamic operating characteristics with regard to electrical transient effects and electromechanical behaviour. The performance of a synchronous machine following a sudden electrical short circuit has already been examined in detail; that of an induction machine differs significantly. Having no specific excitation winding, the induction machine exhibits neither the second component of short circuit, known as the 'transient current', nor the steady-state component, as following a sudden short circuit the current will decay to virtual zero. This leaves only a response similar to the subtransient component of fault current, which has a fast rate of decay.

In the past the contribution of induction machines to system fault level was often ignored for this reason, but with the greatly improved performance of the circuit breakers now in common use this cannot be ignored. Another relevant factor is the effort to improve induction motor efficiency, particularly with large motors, which has resulted in the use of lower values of rotor resistance. This has a direct effect on the decay time of the subtransient current and the increased duration of the fault current contributed by the induction machine must be considered when assessing system fault level.

The electromechanical behaviour of the two types of machine is quite different, as indicated by Figures A1.5 and A1.6. Any variation of the load angle of a synchronous machine (i.e. the displacement between the machine rotor and the constant synchronous speed) results in a transfer of energy between the electrical system and the mechanical rotating system (the prime mover of a generator or the

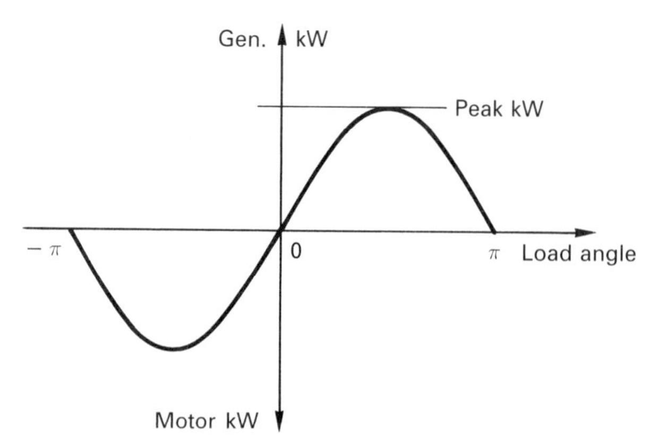

Figure A1.5 *Power–load angle relationship of a synchronous machine*

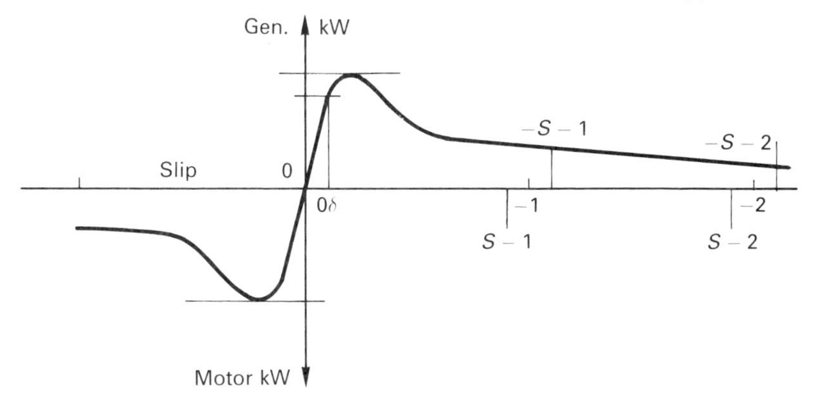

Figure A1.6 *Power–slip relationship of an induction machine*

driven load of a motor). With the induction machine, a corresponding transfer of power requires a slip or speed difference between the rotor and the constant synchronous speed. The significance of this difference is obvious when the equations of motion of the mechanical system are considered, indicating that the synchronous machine is a much 'stiffer' electromechanical unit than an induction machine and reacts much more rapidly to transient changes. In some situations this performance is an advantage, but there are other applications where the behaviour of the induction machine is preferable.

This characteristic is important when selecting the optimum governing system for generator prime movers.

Excitation

When suitably excited, a synchronous generator can produce an electric voltage at any practical speed at the corresponding frequency, and is thus capable of providing power at any speed and gives the possibility of providing a variable-frequency power source which is often desirable for some two-machine systems, such as ship propulsion. It also enables a synchronous generator to supply a system which has no alternative power source; that is, it can operate in an island mode.

If the amount of excitation on a synchronous generator is controlled it can supply a power system at constant voltage or it can be used to give a supply at constant volts per hertz – a condition giving optimum performance for loads such as induction motors, which require operation at constant magnetic flux, when the prime mover speed varies significantly. Such control also enables a generator to supply loads having a wide range of power factors, since it can provide a variable amount of kVAR at any kW load. Fine adjustment of the generator voltage, together with suitable frequency control, enables a synchronous generator to be synchronized with an existing supply system in the conventional manner.

A simple induction generator is unable to perform in this manner, since it has to have an external source of kVAR to operate. When there is a sufficiently large electrical system which will maintain constant frequency and voltage irrespective of the kVAR demand, an induction machine can operate satisfactorily. It will transfer energy from the rotor to the stator or vice versa, depending on the slip and if an induction generator is driven by a prime mover at a negative slip (i.e. the rotor speed exceeds the synchronous speed), it will generate kW into the system. Any variation of prime mover speed will result in a change in the power generated. However, such a machine will not generate an effective voltage until after it has been connected to the supply system, and it cannot be synchronized in the same way as a synchronous generator. It is customary to run an induction generator up to synchronous speed using its prime mover and then to connect it to the supply system, when there will be a transient magnetizing pulse of current drawn from the supply, which will rapidly decay to the current corresponding to the value of steady kVAR required to magnetize the generator.

It is possible to improve the capabilities of an induction generator to overcome some of its disadvantages for specific applications where its other features may be distinctly advantageous. One such device is the provision of capacitance connected to the generator terminals in the form of static capacitors. When such a machine is run at speed the slight remanence in the cores, due to residual magnetic field, can produce a small voltage and this then circulates kVAR through the capacitors. This kVAR can 'excite' the induction generator to produce a higher value of voltage, and the sequence continues until a stable condition exists, when the generated voltage produces a capacitive kVAR equal to that required to produce the voltage. Figure A1.7 shows a typical induction generator saturation curve and indicates how different values of capacitance will produce different generated voltage. By selecting a capacitance to produce rated voltage, conventional synchronizing can be provided for an induction generator. However, if a larger value of capacitance

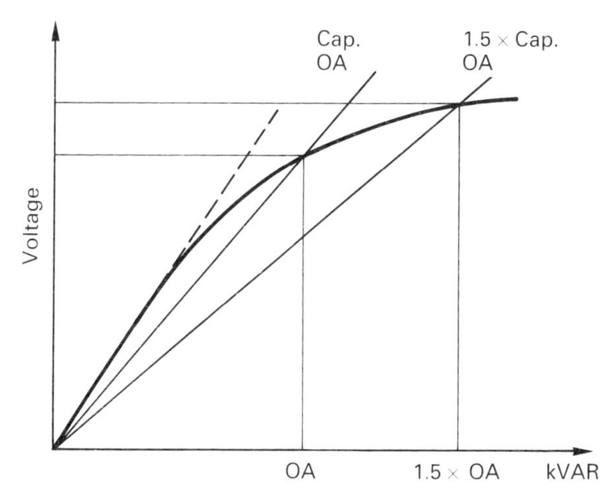

Figure A1.7 *Capacitor excitation of an induction machine*

is used, the voltage generated will exceed the rated value and this can be dangerous, as the voltage could be sufficient to cause damage to equipment connected to the terminals. This phenomenon, known as 'self-excitation of induction machines', must always be examined in detail when using capacitors in conjunction with induction machines in any way.

Figure A1.8 shows the effect of adding capacitance to match the rated voltage-magnetizing kVAR of a generator. Although no kVAR is drawn from the system at no load, an induction generator still

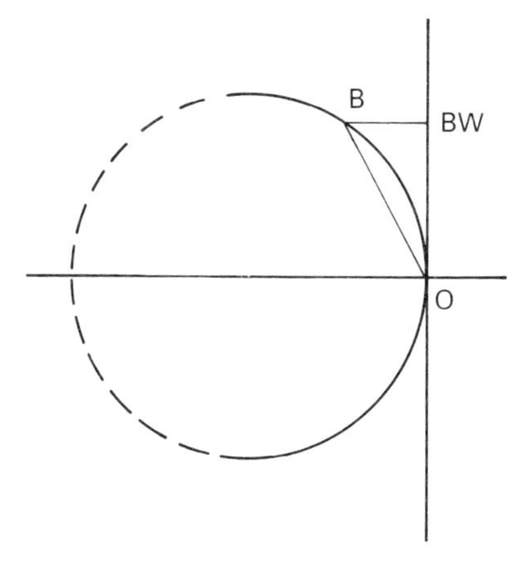

Figure A1.8 *Induction machine kW–kVAR diagram using capacitor compensation*

continues to operate at a lagging power factor for all other values of load. Other synchronous generators on the system will then have to provide this kVAR by increasing their excitation currents.

Secondary factors

The above consideration has assumed ideal operating characteristics and has ignored many secondary effects which can, in particular situations, produce adverse effects. Some of these and their adverse effects are described below, but it is important to remember that these factors all interact and it is their cumulative effects that are important. Many other factors resulting from practical manufacturing and erection conditions are relevant but are not of sufficient importance to be considered here.

Unbalanced and harmonic loads

If the electrical load is unbalanced between phases, then it is possible to represent it as the sum of three component loads, each symmetrical between phases and each producing an electromagnetic field: one which rotates at normal synchronous speed and is called the 'positive phase sequence' component; another which rotates at synchronous speed, but in reverse direction, and is called the 'negative phase sequence' component; and the third, which remains stationary in space and is known as the 'zero phase sequence' component.

The positive phase sequence component will behave in the manner described for a balanced phase system but the other two components will interact in different electromagnetic ways as their relative speeds with the rotor will differ from that of the positive sequence. With a synchronous machine the positive sequence rotates at a speed equal to that of the rotor, the zero sequence at a speed equal to synchronous speed backwards relative to the rotor, and the negative sequence at a speed equal to twice the synchronous speed backwards. With an induction motor operating at a slip of S relative to the positive sequence speed, the slip of the rotor with respect to the zero sequence will be $S - 1$, and with respect to the negative sequence will be $S - 2$. The corresponding values for an induction generator with a slip of $-S$ will be $-S - 1$ and $-S - 2$; reference to Figure A1.6 shows that under these conditions the machine is in the generating mode. However, the frequencies of current associated with these two components are higher than normal and this results in increased losses and consequent additional heating in the windings, particularly on the rotor. It is therefore desirable to keep system unbalance to a minimum if the machine is required to operate at its rated load and operating conditions of temperature rise and efficiency etc.

Loads which contain harmonic currents or voltages have similar effects on machines, as each individual harmonic component behaves in a similar manner to the unbalanced phase-sequence components and can produce motoring or generating power as well as additional losses and heating.

All such load distortions result in a deterioration of the ideal fundamental performance capability of either type of machine; reduction in operating efficiency and increase in winding heating are usually the most serious results.

Practical machine features

All electrical windings possess resistance and reactance and when carrying current these form significant kW and kVAR loads which must be provided from the electrical system or the machine itself. All the simple diagrams, Figures A1.1 to A1.8, would be modified to some extent if these factors were included and the apparent symmetry between the motoring and generating characteristics would be found to be incorrect. In normal operation these effects are not very significant, but under unusual conditions, such as operation at low frequency, they become important and it is essential that they are included in all machine design and operation analysis.

The saliency effects produced by the salient poles on a synchronous machine are significant under steady-state normal conditions and differ appreciably from the effects of the round rotor machine, as has been discussed previously. This represents an added advantage of the salient pole generator.

The magnetic components of all machines exhibit saturation effects, as indicated in Figure A1.7: that is, the relationship between winding voltage and magnetizing kVAR is nonlinear, and in fact becomes asymptotic. This naturally has a serious effect on the electromagnetic behaviour of the machine fields, as discussed previously, and can have limiting effects on high-load conditions, particularly in the region of the pull-out point or with induction machines at low speeds where their reactance is reduced and high values of kVAR may be circulating. Again, an accurate assessment of this factor must be made when specific characteristics are required under such extreme operating conditions.

Choice of synchronous or induction generators

There is a great difference between these two types of machine both during normal steady-state operating conditions and also during abnormal and transient conditions, when all the practical design features are considered. In many special applications these secondary factors can be the deciding points in selecting the most suitable type of generator. However, in the great majority of instances it is not necessary to consider all the features described, as there are quite wide ranges of applications for which each type of machine has distinct fundamental advantages. The two major aspects relevant to this choice are (1) the prime mover being provided to drive the generator and (2) the electrical load to be supplied and the system in which it is connected. The significance of these have already been discussed in detail for the synchronous generator, since this type comprises the vast proportion of generators in service.

Island mode (two-machine)

A generator operating in the island mode – i.e. unconnected to another source of power – determines the system frequency and voltage. The synchronous machine is ideal, since the frequency is proportional to prime mover speed and the accuracy of the governor/fuel system determines the accuracy of the system frequency. When provided with an adequate excitation system and automatic regulator, the synchronous generator will maintain constant system voltage with any value of load kW and kVAR within its designed operating range. An induction generator in this mode does not produce a frequency proportional to the prime mover speed, as the slip difference varies with the load being supplied. The generated voltage would depend on a supply of capacitative kVAR and this would require to be infinitely variable to deal with all possible load conditions: this is generally impracticable. Thus the induction generator has no real advantage in this mode.

Island mode

This condition is similar to the two-machine mode in that the generator is not connected to a large generation system which would maintain a constant frequency and a closely regulated system voltage. Such a generator may operate on its own, in which case the behaviour is similar to that of the two-machine mode except that the load will be a distributed, mixed one in its own local distribution network. The advantages of the synchronous generator remain as before, but an induction generator can have advantages if used together with a synchronous generator capable of supplying the kVAR required to magnetize the induction machine as well as supplying the load kVAR. Thus the induction generator merely functions to provide MW to the system, and its lower first cost may justify its use. It has secondary advantages in that its construction is usually simpler and more basic than that of the synchronous machine and service and maintenance should be less. The relationship between generated kW and rotor speed enables such a machine to function with a prime mover that has a less accurate governing system, since it acts as a variable kinetic energy between the input mechanical power swings within its own pull-out limitations. However, it is essential that the induction generator is never left connected to the system without a suitable magnetizing kVAR source.

An island mode system with synchronous generators has the advantage that it can be connected to a larger system using conventional synchronizing procedures for the complete subsystem, and it can then be classed as a satellite mode system.

Satellite mode

Operation in the satellite mode involves a local system which is similar to an island mode system, consisting of one or more generators and a distributed local load, connected to a much larger system through a relatively high impedance link. In this mode all the advantages described for the island mode still apply, but the synchronous generator now has an added advantage in that it can be used to compensate for the impedance of the interconnecting link and maintain a stable island voltage. Without such a generator and control function, the island site would have a significant voltage drop as the load increased.

Large system

A large system can be regarded as one which has enough generating capability and suitable control functions to maintain constant frequency and voltage irrespective of a new generator being added. This implies that, in the case of an induction generator, an adequate supply of magnetizing kVAR will be available without affecting the terminal voltage. These conditions exist for many waste energy recovery schemes which can use a basic form of prime mover to provide kW power to the electrical system and can show considerable economic advantages when used in conjunction with induction generators. There is no requirement for close speed regulation and variations in fuel quality or supply are not inconvenient. No sophisticated control systems are necessary and the relative simplicity and cheapness of the induction generator render it attractive for this purpose.

In every application, of course, it is still necessary to consider the other factors described to determine whether special conditions exist that indicate an overriding preference for one or other type of generator. Among these the most usual are the effect of the new generator (1) on the fault rating of the local system and installed equipment, (2) on the transient recovery behaviour of the local electrical system and (3) on the electromechanical effects within the generating set itself, involving the inherent characteristics of the electrical system load and the prime mover.

Application

The most common type of electrical machine in service is the induction motor, of which the range of types and sizes is very wide. As a result of mass production the cost of some machines is very low when compared with other types of machine of the same size. Since any induction motor can be used as a generator, it is often assumed that all such machines are suitable for this purpose. This is not correct, since the electrical characteristics required for a motor and its mechanical construction may not be suitable for a particular prime mover.

A standard induction motor will not normally incorporate every feature that would normally be provided with a machine specifically designed for generator operation. However, if its shortcomings are not too limiting, its lower first cost could make it a satisfactory economic choice in preference to a more expensive machine designed to meet all criteria.

The electrical parameters which are of prime importance are the slip at rated kW and the magnitude of its peak kW and the slip at which it occurs. Motors with higher efficiency tend to have smaller values of slip and hence a steeper kW–slip characteristic. This results in a larger fluctuation in generated kW resulting from inaccuracies in the speed control of the prime mover. Usually a higher-slip machine is advantageous for an induction generator acting as a base machine, i.e. providing a fixed amount of power to the system equal to the output of the prime mover.

The peak kW determines the maximum power that can be accepted from the prime mover; if it attempted to provide more it would accelerate the induction generator above its peak kW slip, when the electrically generated power would start to decrease and the set would continue to accelerate until some control action reduced the prime mover energy input.

The greater the slip for peak kW the greater the stability margin of the set, since it can absorb more acceleration power before running away and permit more time for limiting control action to be taken.

The peak kW is related to the generator impedance. A machine that is designed for a higher value than standard for a motor will have a corresponding lower impedance and this can have disadvantages from the point of view of contribution to the system fault level, as the generator contribution is related to this impedance value. Thus the benefits of higher transient stability have to be assessed in relation to the increase of fault level contribution, and a compromise adopted depending on the more significant of the two features.

Induction motors may be designed to give specific starting conditions for particular applications. A wide range of characteristics can be obtained by adjusting the machine reactance and selecting appropriate starting winding design, incorporating features such as 'deep bars', 'shaped bars' and double or triple windings. This aspect of design is important as it affects the subtransient behaviour of a machine when acting as a generator, both in respect to reactance and resistance, and directly affects its contribution to an electrical system fault. In general it is desirable to keep this contribution as low as possible and this usually conflicts with the use of a motor, with enhanced starting characteristics, as a generator.

The other characteristic which may be of prime importance is the kVAR demand of the induction motor. This tends to increase with machine reactance and also with radial length of air-gap, and these tend to be kept to a minimum consistent with ability to meet the standard requirements for an induction motor. However, when an induction generator is being used on a large system the importance of reducing kVAR demand is not so great, and other aspects of these features can be considered.

Many induction motors are similar in construction. Most driven units are provided with their own shaft support systems and motors, which are also self-contained and possess two bearings, which can easily be flexibly coupled without presenting severe mechanical problems. With this arrangement the motor can have a very small air-gap, as effects of misalignment of rotor and stator and of unbalanced magnetic pull are greatly reduced. However, this does not always apply when using the machine as a generator connected to a prime mover and this factor must be carefully assessed when determining a suitable design for a generator. Whenever possible a larger air-gap should be provided if accurate alignment cannot be guaranteed under all possible operating conditions.

The possibility of runaway overspeed by a prime mover should also be considered and it may be necessary to design the generator for a greater value of overspeed than is provided for the equivalent induction motor.

The other phenomenon which can present serious problems with induction machines is self-excitation, as previously described. Such a machine, with a source of mechanical power such as a prime mover, when left with its windings connected to a value of capacitance greater than its own no-load kVAR, can self-excite and, depending on the kVAR relative values, can produce overvoltages possibly dangerous to the machine itself and the connected capacitance, and also to any apparatus also connected to the system.

It is most important, therefore, to ensure that in any unusual operating conditions an induction generator cannot be subjected to excessive overexcitation.

Appendix 2
Protection

Protection equipment is normally regarded as that provided to reduce the consequences of an electrical failure or fault. It is often based on the behaviour of plant operating on a large system, and is sometimes designed to be suitable only for such conditions. This enables protection schemes to be developed for specific sets of standard conditions related to a fault and these are often treated as independent non-interactive systems dealing with specific electrical units such as generators, transformers, distribution, motors etc.

When dealing with small or island systems this philosophy is no longer relevant. The unique operating conditions, and wider range of applications, must be considered when selecting suitable protection equipment. Such equipment can be used to reduce the consequences of a fault not only to the electrical equipment itself but also to associated mechanical plant, personnel, and sometimes to the complete installation such as an offshore platform or a ship. In each application the respective priorities have to be established precisely, as it may be necessary to sacrifice some electrical plant in order to ensure the safety of personnel or of some more valuable units of the installation.

In such instances standard protection schemes may be unsuitable and a total integrated protection system must be designed as a package. The effects of what could be a relatively simple fault on a large system may be very complicated on a small system and the interaction of all protective functions must be examined for each condition.

Single function relays

Protection equipment operates by monitoring a function related to the performance of the system; relating this to some reference value can determine whether an abnormality exists. The significance of this abnormality has then to be related to some action which should be taken. This apparently abstract procedure is simplified in general by making generalizations or assumptions such that simple criteria can be applied to single functions and standard actions taken if the criterion is exceeded.

Thus, in electrical equipment, windings have two distinct criteria: the current they can carry without sustaining damage, and the voltage they can withstand without damage to their insulation. This would suggest that a simple measurement of current and voltage would indicate when these exceeded the design rated values; the winding could then be disconnected from the supply. Such simple function protection, however, is impractical since the damage/failure mechanism in the equipment is not dependent solely upon the instantaneous value of current or voltage.

One very significant factor involved is time, and a second, equally important, is the environment, since the plant limitation is usually determined by a combination of electrical, thermal, dynamic and chemical effects. The temperature attained by the insulation adjacent to the conductor is determined by the relation of energy flow to the point provided by the integral of current squared and time, and the dissipation of this energy either along the conductor or through the insulation to the outside cooling media. Thus the temperature, thermal characteristics and rate of flow of the coolant will affect the temperature through the insulation, independent of the current. A simple current value will not define a specific insulation temperature.

In practice, therefore, it is customary to select standard operating conditions which are used to demonstrate the capability of the plant, by factory testing if necessary, and to use these as safe operating criteria on the assumption that, in practice, the nett effects of deviation in actual service never cause the plant to exceed its rating. This then allows a simple current measurement to be used to check that the steady load is within the plant capability.

However, many electrical loads are not constant and some vary widely and frequently, which invalidates this simple criterion. Two alternative approaches are in common use. One is to measure the temperature on the outside of the coil insulation and deduce from this whether the inside temperature is within rating. The other is to use a simple thermal model of the winding and, by passing a proportional current through this, determine when the model, and therefore the main equipment, has reached its thermal limits.

Initially, a simple inverse relationship was assumed for this latter function and relays incorporating inverse time characteristics are in universal use for this purpose. These can now be obtained with optional characteristics, the better to match the actual characteristics of the equipment which they are used to protect. More sophisticated relays are now available with even more elaborate model simulations, but the difficulty lies in obtaining accurate data on the actual equipment characteristics to enable the correct model to be selected.

Typical values for multirange overcurrent relays are:

Standard inverse time $\dfrac{0.14}{I^{0.02} - 1}$ seconds

Highly inverse time $\dfrac{13.5}{I - 1}$ seconds

Extremely inverse time $\dfrac{80}{I^2 - 1}$ seconds

In addition to thermal effects, winding current can produce damage due to mechanical forces if the magnitude is too great. At even greater values of current the conductor may melt as a consequence of inability to remove the heat being generated. These higher values of current can usually only be encountered following a short circuit within or adjacent to the winding and occur over a very short time interval. It is usual to provide a separate device to detect such dangerous magnitudes of current; and it is assumed that their occurrence is indication of a dangerous system fault and action is taken to minimize damage by disconnecting the winding from the supply. In some instances fault currents may not reach the full short circuit prospective value if there is significant fault impedance and it is usual to adjust the inverse time–current device to detect such currents and take the necessary clearance action.

Steady overvoltage on a system usually indicates failure of some control device such as an AVR or tap-changer and is more usually associated with such devices rather than being applied to measuring overvoltage on specific windings. However, short-term overvoltages can be produced during transient conditions such as circuit interruption or closing, or can be induced by lightning surges. Again these devices are usually applied as system protective devices rather than applied to separate windings, although it may be necessary to provide back-up protection, e.g. surge suppressors on machines in addition to system diverters, to ensure a better level of protection on equipment having a lower insulation impulse ratio than that of the main system devices.

Undervoltage does not impose any risk on a winding but it can have adverse effects on the characteristics of a piece of equipment and can result in its failure to perform its duty. Protection can be provided by isolating such equipment but in many cases the undervoltage is due to a transient condition and after a short time it can recover to near normal. A protective device which operates solely on undervoltage would trip the equipment, probably unnecessarily, since if not disconnected it would probably return to normal operation on restoration of the system voltage. For this reason any undervoltage device related to a drive has a time delay characteristic to permit such recovery; it will only isolate it if the undervoltage persists.

Compound function relays

It will be seen that single function protective devices have only limited application if accurate operation is required and that they are only capable of simple levels of protection of single items of equipment such as a winding. In practice all equipments have more complicated electrical systems and require the use of compound function relays to detect either risks to the plant or the consequences of a fault.

Most power system devices are three-phase and although all phase voltage and currents should normally be balanced this condition is not always met. It is therefore necessary to use three single function relays to measure any deviation in current or voltage above the setting value. In addition to the effects of a single value exceeding its rated value, phase unbalance produces other adverse effects associated with the negative- or zero-phase sequence components produced. These are not numerically equal to the magnitude of phase unbalance, but a simple detection of such a condition is obtained by comparing the three single-phase values and detecting when the deviation exceeds a preset value. This can be done quite simply in a relay incorporating direct measurement of each phase value and this comprises a compound relay detecting overload and unbalance. When the plant involved is particularly sensitive to negative phase sequence effects, a compound relay can

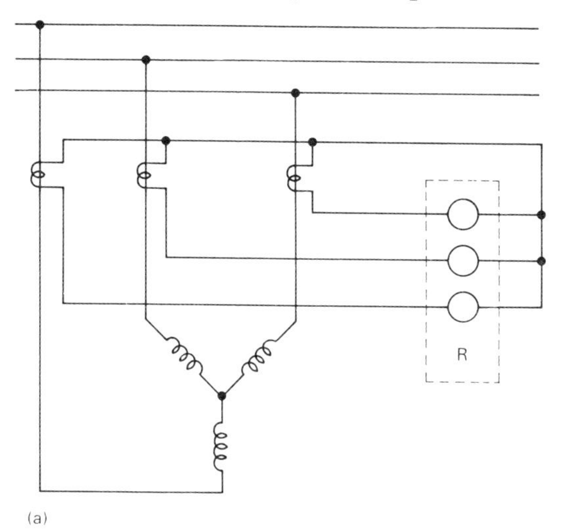

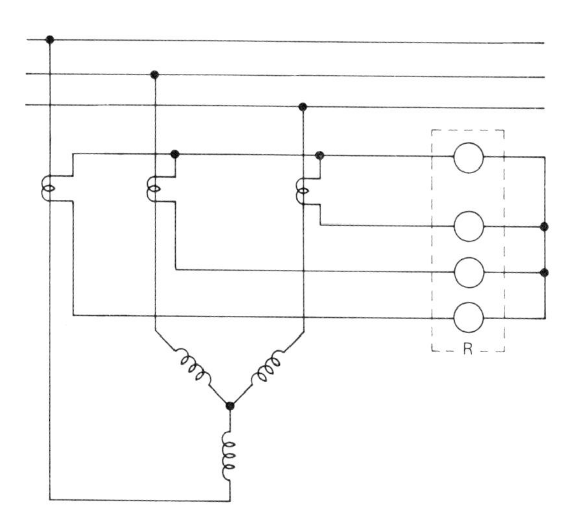

Figure A2.1 *(a) Three-phase overcurrent. (b) Three-phase overcurrent plus earth fault – no neutral available*

be used which measures the three-phase currents and then deduces from this the three magnitudes of the positive-, negative- and zero-phase components and can operate at preset values of each of these (Figure A2.1). By measuring both current and voltage and the phase angle between them it is possible for a relay to evaluate the corresponding values of MW and MVAR and operate at any preselected value. Measurement can be carried out on a single phase which can be used for total power if all phases are symmetrical, but if any unbalance is expected an accurate measurement requires a full three-phase relay measuring all phase voltages, currents and phase angles. As the same input data are required for both MW and MVAR it is convenient to feed the two relays from the same voltage and current transformers.

Voltage-restrained relays are available which are used to give overcurrent protection, typically on generators, where the significance of the current flowing is affected by the system voltage existing: at normal operation and rated voltage the relay has one definite current–time characteristic but, should a fault occur which reduces the system voltage, the relay functions on a different current–time curve.

Reverse power

Electrical sensing devices can be useful to protect associated mechanical equipment, as they provide accurate data on loading very quickly and can be used to take remedial measures if dangerous conditions are detected.

Mechanical prime movers driving generators operate satisfactorily when producing power up to their rated overload conditions, but should conditions arise, e.g. interruption of fuel supply, in which they can no longer produce positive torque, the coupled generator will immediately operate as an electric motor and try to maintain the set at full speed. Depending upon the type of prime mover, it will absorb power from the electrical system via the generator, and dissipation of this can produce adverse conditions in the machine which could cause damage. With prime movers liable to suffer such damage it is usual to provide reverse power protection on the driven generator; this can detect when the power flow from prime mover to generator is reversed. In service this condition can arise during transient conditions such as significant changes of load on the system or when a generator is being synchronized with the system. It is necessary, therefore, to impose some qualification on the reverse power to avoid unnecessary operation and this is usually provided by some form of time control. This is set at a time which will be longer than any power reversal that can occur under healthy conditions but less than the time likely to result in damage to the prime mover if the faulty condition continues to exist. In addition, prime movers can carry a limited amount of power and the value of reverse power at which the device will operate is adjustable so that it will only operate when a dangerous condition is known to exist. This protection for reverse power usually takes the form of an adjustable setting of reverse power level with a delay time before action is implemented.

Power reversal can also occur on drives involving electric motors, since any reduction in system frequency will enable energy stored in the motor-driven unit to be converted by the motor into electrical power acting as a generator. With small system disturbances this effect usually has negligible effect on the plant and can be ignored, but in the event of serious disturbance resulting in considerable frequency drop, the amount of reverse power flowing can be considerable. Electric drives which require serious consideration are those where the driven member operates into a high-energy main such as a high-lift pump installation, when loss of positive torque from the motor enables the stored hydraulic energy to convert the pump into a turbine form of prime mover and drive the motor as a generator. Under this condition it is preferable not to trip the motor from the system; otherwise all the stored energy in the hydraulic system will be converted into kinetic energy of the set and result in an uncontrolled overspeed. It is preferable to leave the motor connected and generating electrical power into the system, retaining the set at about full speed. This it will do up to its pull-out torque limit, and this parameter should always be checked for suitability on drives where this condition is likely to occur.

Thus the application of reverse power protection should always be carefully assessed for each installation and not automatically included as a matter of course.

Frequency

Another function on a system which can produce adverse effects is the frequency. On large systems frequency variation is usually negligible, but on small and island systems it can be serious. The direct effect of frequency variation is on the speed of electrical drives; this will usually affect their characteristics and change their power demand. The drive characteristics may cause serious disturbance to the plant process and this factor should be considered when determining what protective devices are used and what action they should initiate. Thus a centrifugal pump which has a torque demand proportional to the square of the speed could require a MW demand varying as the cube of the frequency. The change in pressure head of the pump could also result in failure of the associated hydraulic system.

In addition, frequency variation has adverse effects on induction drive motors where the core losses and friction and windage increase rapidly with increase in frequency. The flux in such motors is proportional to the terminal voltage and inversely to the frequency. The torque produced is consequently reduced if the voltage decreases or the frequency increases and its ability to maintain the drive can be impaired. If the values of voltage and frequency vary simultaneously, provided that the volts-per-cycle value remains constant the motor flux and hence torque will be maintained; however, the speed will still change. In some two-machine systems it is convenient to utilize a constant volts-per-cycle voltage regulator which will ensure optimum operation of both motors and generators, even when operating over a range of frequency and speeds.

In general, AVRs now include a volts-per-cycle protection feature to prevent generators being overfluxed should the AVR be operational when the prime mover is running the generator up to speed from standstill or down from full speed.

A system frequency disturbance occurs when the power balance is altered either by change in load or in prime-mover-driven generator power output. Such changes can occur normally in a system, but if the governor control system is functioning effectively it will compensate for any changes to restore the preset system frequency within a very short time. Only a negligible frequency excursion will be observed. However, if the magnitude of the load change is great and occurs rapidly, a significant transient frequency excursion will be encountered and the time to recover to normal will depend on the magnitude of the dip and the characteristics of the governing systems, together with the electrodynamic parameters of the system.

The most common source of serious frequency disturbance, however, is plant failure or a system electrical fault. Plant failure usually results in a sudden loss of load or generator output, with the results indicated above. However, an electrical fault can have much more complex consequences.

When a system fault occurs it causes a sudden change in impedance locally, resulting in a severe reduction in the voltage which, in turn, causes a reduction in the MW demand on the generators. There will then be a sudden increase in prime mover/generator set speed and system frequency, until the governing system can operate to reduce fuel to slow the sets or until the protection system can detect the fault and successfully isolate it. Depending on how the isolation is performed, the system may remain viable, with residual plant capable of operating normally when steady conditions have been restored, or an unstable system

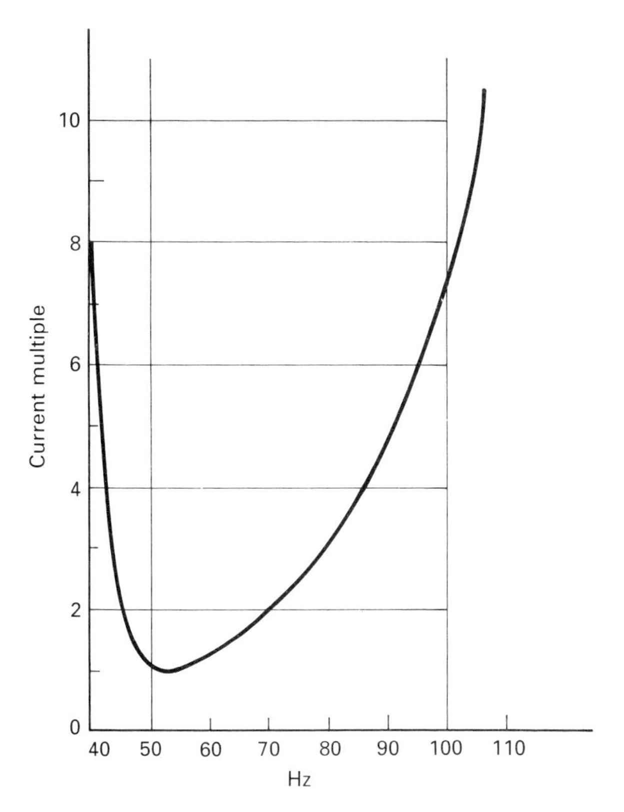

Figure A2.2 *Frequency response of a tuned relay*

may be left which will be unable to recover unless some supplementary control actions are taken sufficiently rapidly. Following clearance of the fault system, load will increase as the voltage recovers and this will tend to reduce the generator set speed and system frequency and activate the governing system to increase fuel to restore normal conditions.

It will be appreciated that frequency alone cannot be used as a positive criterion for the optimum action to be taken following any system fault condition. It is also necessary to determine the simultaneous behaviour of system voltage, and to decide what effect isolation of the fault will have on the residual system.

Frequency variations also affect the operating characteristics of relays and can change their tripping values significantly. It is necessary, therefore, to use relays which will give satisfactory operation over the range of frequency encountered in service. Relays such as overcurrent devices, which are intended to operate on a single parameter irrespective of actual system frequency,

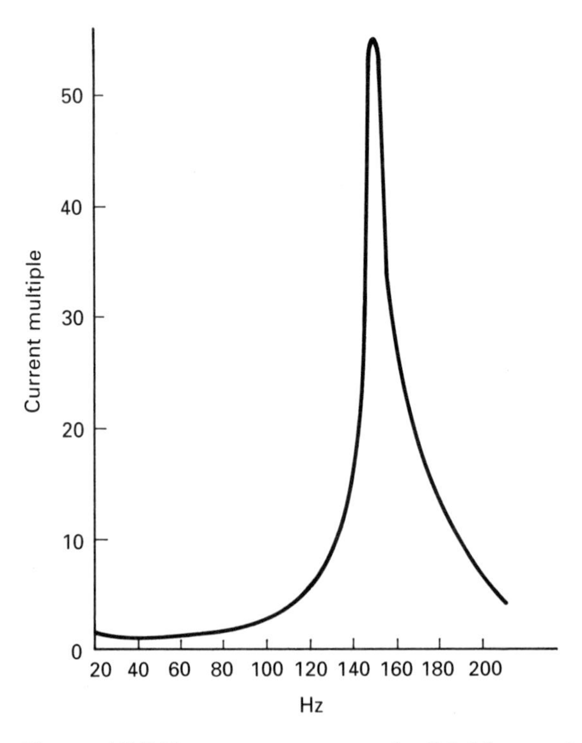

Figure A2.3 *Frequency response of a third harmonic rejection relay*

require a flat frequency response curve. Others, which are intended to operate at the fundamental power frequency, 50–60 Hz, can have a characteristic such as that shown in Figure A2.2 which rejects higher-order frequencies; yet others, which are required to reject a particular frequency such as the third harmonic, can have a characteristic such as that shown in Figure A2.3.

Fault detection

One of the most important aspects of protection is to detect rapidly and accurately any dangerous fault condition on the system. In an electrical system the most dangerous faults are usually adjacent to

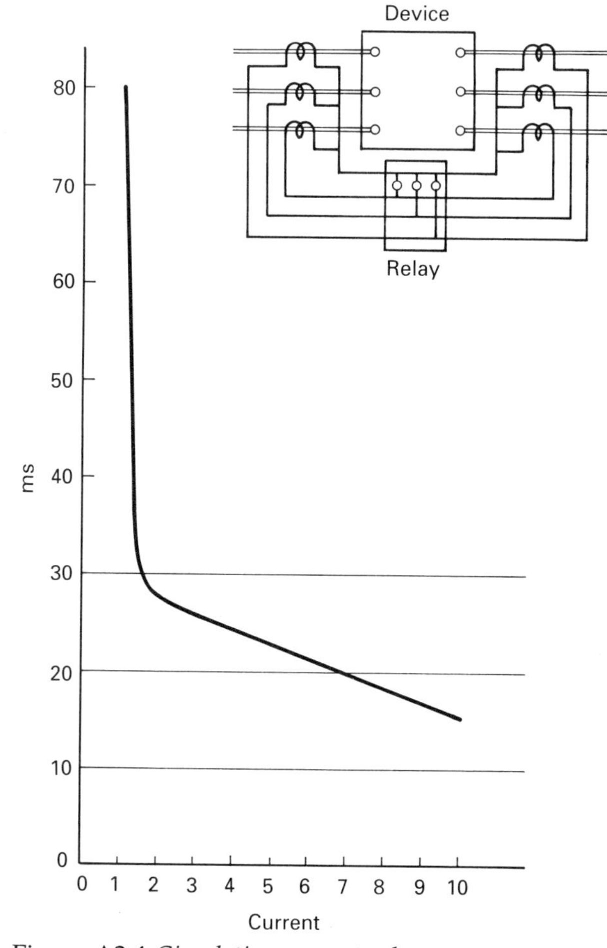

Figure A2.4 *Circulating current relay*

the generators, normally at the high-voltage busbars on the main switchboard. At this location the fault level is greatest and fault damage most severe, and any fault has its most directly adverse effect on system generation capacity. Downstream from this location the impedance increases and the effects of faults are less severe. Protection is graded downstream such that only that part of the system including the fault is disconnected from the generation, leaving as much of the system as possible available for recovery and continued operation. Such discrimination is most easily provided by inverse time–current relays but this is not always practicable through the whole network on small systems. Differential systems, which can be arranged to identify the zone in which a fault occurs, can be designed for very fast operation as they do not need to be incorporated into the discrimination sequence. The zones must be carefully selected so that the specific effects of the fault can be predetermined positively and all the desirable consequent actions determined to result in system stability (Figures A2.4 and A2.5).

Conventional devices can be adopted for fault detection but the manner in which they operate requires careful consideration, usually in conjunction with the results of a compatible system transient stability study incorporating fault level evaluation and load flows with the various residual system configurations which will result from protection operation actions.

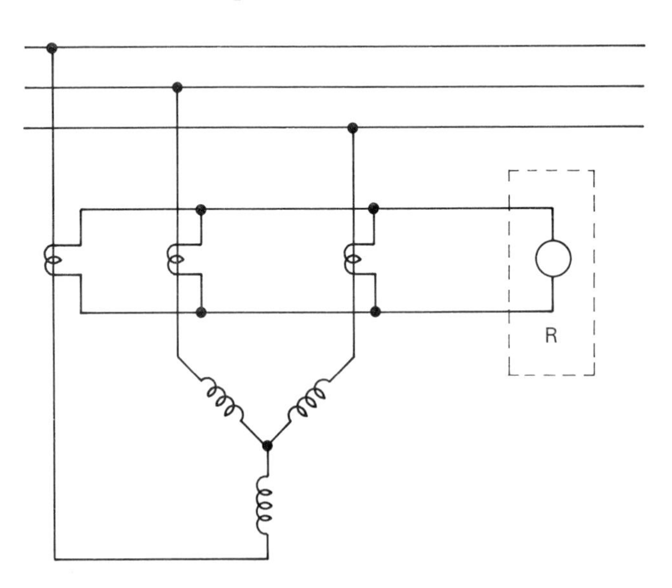

Figure A2.5 *Differential relay with no neutral available*

System earthing

It is now customary to earth all power systems in the interests of safety but there are still a few special conditions where a nominally unearthed system can have advantages. Even such a system is connected to earth by its capacitance and also its leakage resistance, and allowance must be made for these values when investigating the protection required and safety of the system (Figure A2.6). The impedance to earth of a nominally unearthed system can be very variable depending on the amount of network connected and the environmental conditions existing. It is necessary when designing a suitable protection system to have a specific value of impedance to earth, and it is customary to incorporate a specific impedance to earth at a suitable location in the system. This may be of a relatively high, medium or low impedance, with a mixture of resistance and reactance as required for its specific function. In some low-voltage systems the neutral may be earthed solidly but the fault impedance on such systems is usually significant and limits any fault current to a practicable value.

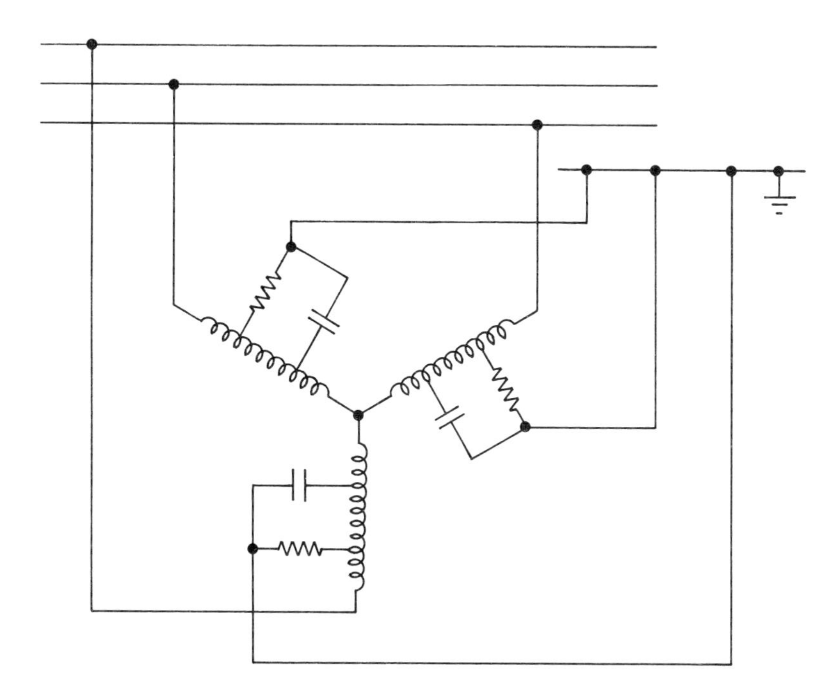

Figure A2.6 *Simplified diagram of an unearthed generator*

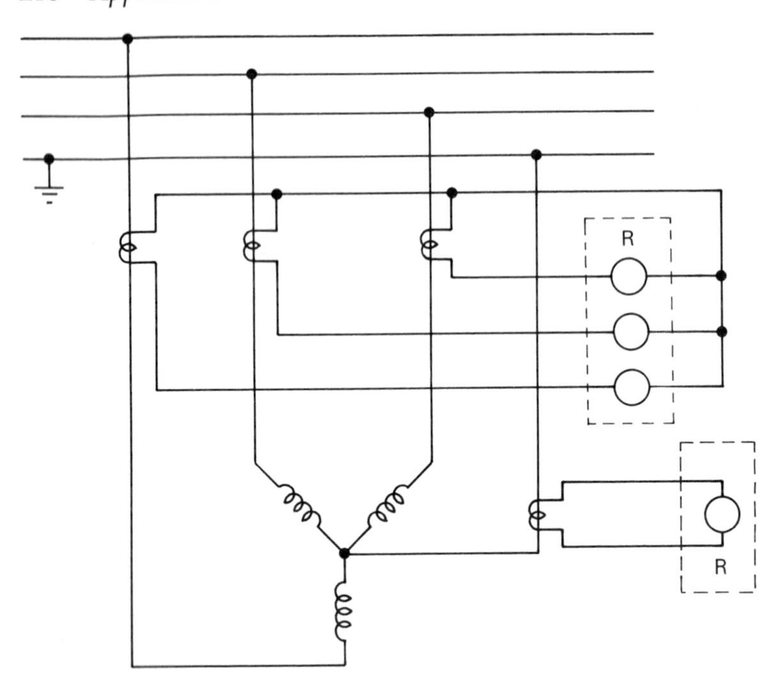

Figure A2.7 *Three-phase overcurent plus separate earth fault, with neutral available*

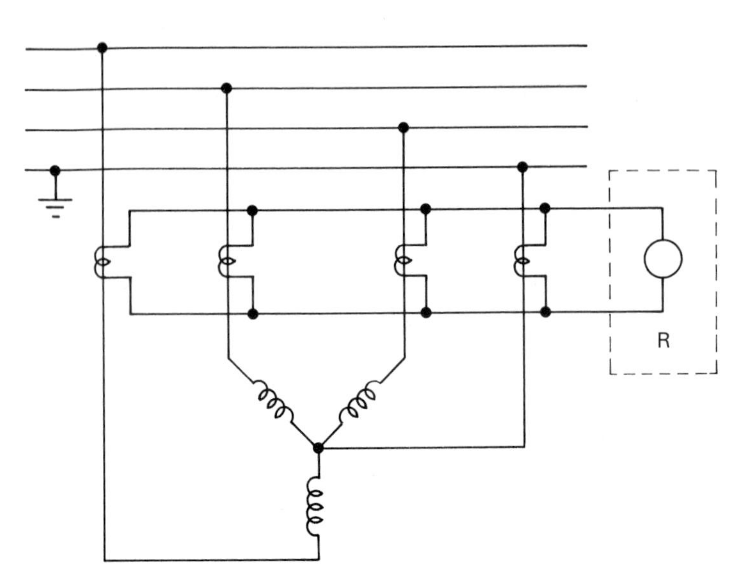

Figure A2.8 *Differential protection with neutral available*

The purpose of a system earth impedance is to provide a positive return path for any earth fault occurring on the system. It serves to limit the voltage to earth attained by any part of the system as well as providing a source of signal to indicate that a fault has occurred, together with a measure of its severity (Figures A2.7 and A2.8). A low-impedance earth permits a large earth fault current to flow and can provide a signal to give rapid isolation of the fault when

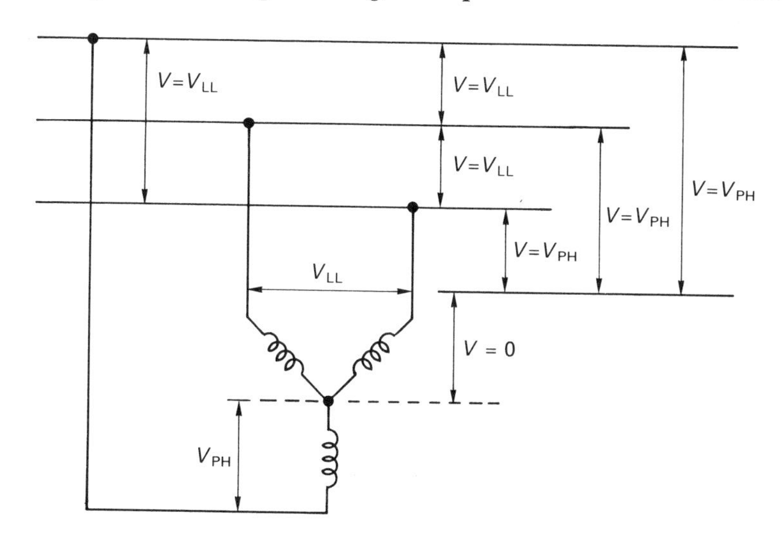

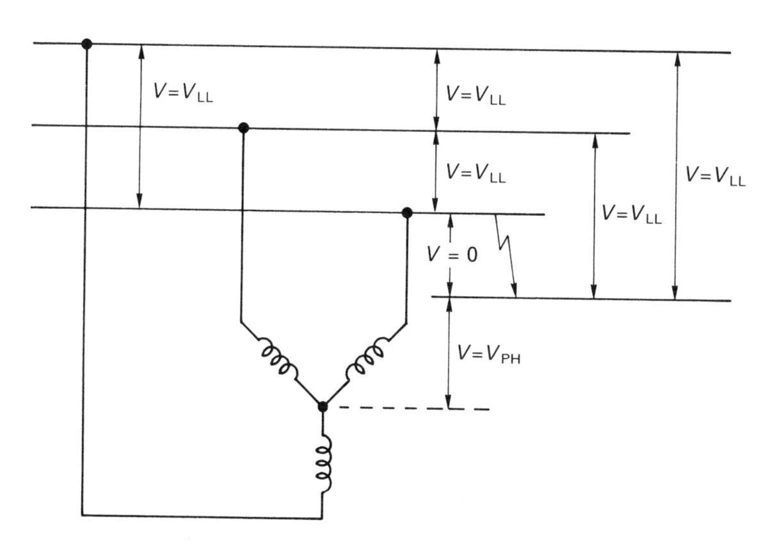

Figure A2.9 *Effect of earthing one line on an unearthed system*

properly located. However, the possibly high value of fault current can result in severe damage, and rapid positive clearance is essential to avoid serious loss of plant. A high-impedance earth limits the magnitude of fault current but, as a result, the neutral is displaced from earth potential and can cause parts of the system to attain a value of line-to-line voltage to earth (Figure A2.9). Most equipment is now designed to be capable of operating under this condition for a considerable period without serious deterioration of its insulation. If the magnitude of the earth fault current is kept below the value likely to cause damage to equipment, the system can continue normally with such an earth fault existing. On certain essential small systems, such as ships and offshore platforms, the availability of full power for a reasonable time following indication that an earth fault has occurred can enable emergency measures to be taken in a safe and proper manner, avoiding hazard to the plant or structure.

Should a second earth fault occur under such conditions (which would be most unlikely), the double fault would require rapid shutdown, since the earth fault current would no longer be determined by the earthing impedance.

As a compromise between these two extremes, earthing such as to limit earth fault current to the order of full-load current of the equipment involved still gives a healthy signal to indicate fault status, and the speed of isolation possible reduces plant damage by this current to a reasonable level.

The method of arranging system earthing varies considerably with the extent of the system involved, the nature of the equipment included and the manner in which it is operated (Figure A2.10).

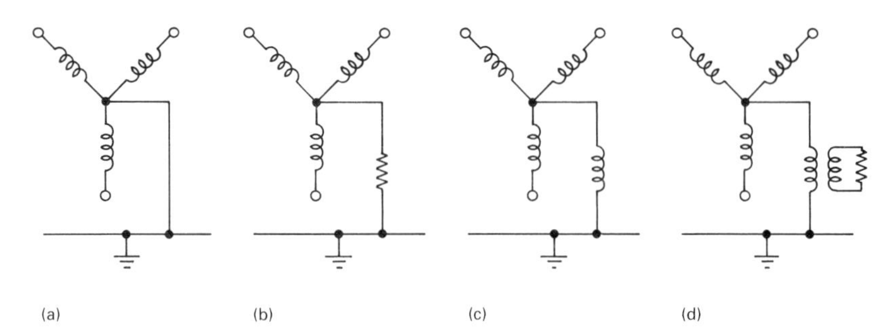

Figure A2.10 *Alternative methods of earthing: (a) solid; (b) resistive; (c) reactive; (d) high voltage*

Where only a single generator is involved, it is logical to connect its neutral to earth. However, when several generators may be operating in parallel on the same busbars it is only necessary to connect one generator neutral to earth, and usually this is done on the largest generator. As this generator cannot always be connected to the busbars it is necessary to make provision for at least one other generator to be available for connection to earth in order to earth the system. In the event of both generator neutrals being connected to earth, the possible earth fault current would be doubled and it is usual with this arrangement to provide isolating switches in the generator neutral connection and interlock these so that only one earthing device is in circuit at any one time. It may be necessary to provide alternative earthing devices on more than two generators to ensure that the system can always be earthed (Figure A2.11).

In addition to increasing the possible earth fault current, multiple earthing allows harmonic currents to circulate between all generators or transformers that are connected to earth. These can produce extra losses and heating in the machine windings and can possibly affect control devices.

Small systems with only a few generators can have a simpler earthing arrangement if high-resistance earthing is used, since multiple earthing is possible. Because of the high earth impedance,

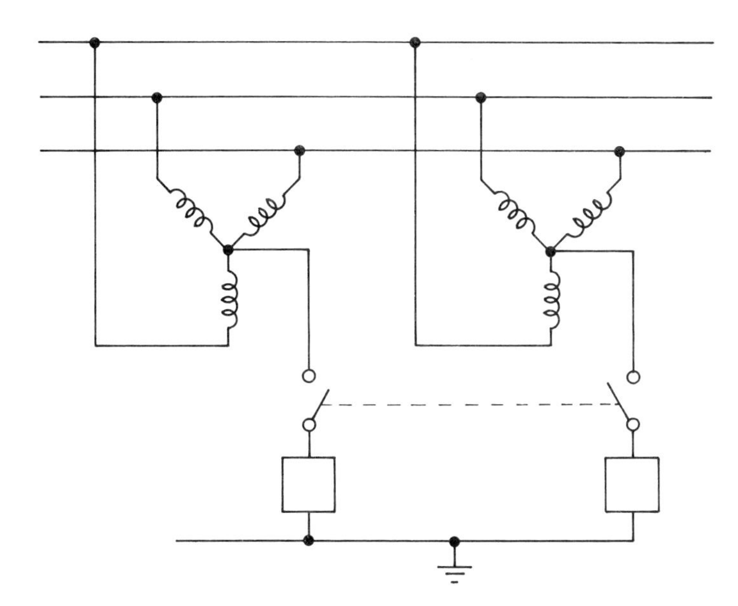

Figure A2.11 *Multiple generator earthing*

circulating harmonic currents are suppressed. By selecting a value of impedance that gives a safe value of sustained fault current with the normal number of generators in use in service, operation with an earth fault can be maintained. When operating with fewer generators, the magnitude of an earth fault current is reduced but, by using sensitive relays, warning of a fault can still be given.

Generator earthing is adequate for the directly-connected system but, where power transformers are used to supply other parts of the system, by connecting these as delta/star the starpoint can be earthed in the same way as a generator and so provide safe earthing on the secondary system (Figure A2.12). Where a convenient starpoint is not available for secondary earthing, an artificial earth can be provided by connecting a distribution transformer with a zig-zag connection to the secondary system.

Earth fault detection can be provided by measuring the current through the earthing device or by measuring the voltage produced across this device. The preference depends on the magnitudes of the values involved and the adverse effects of the harmonics that may be present. Location of fault is most easily obtained by measuring zone current differentially.

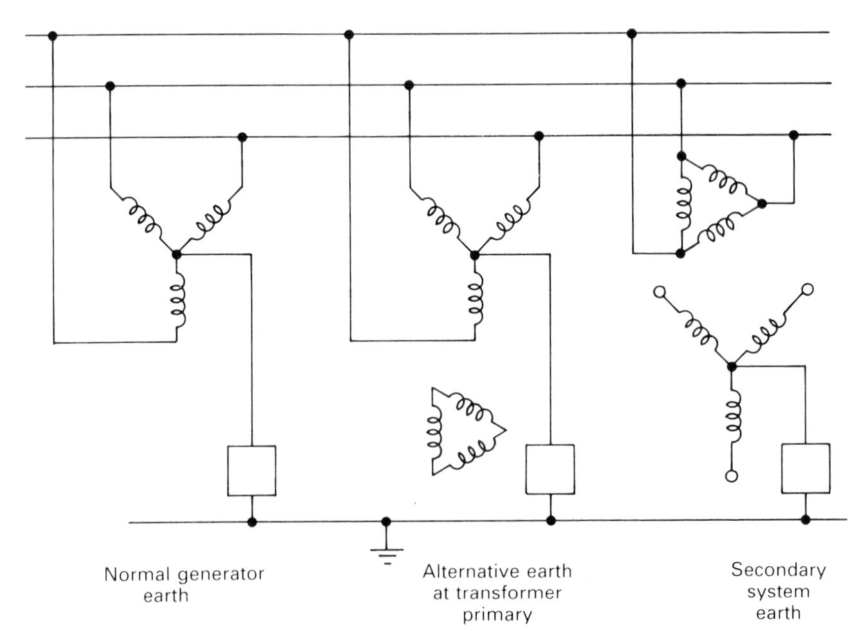

Figure A2.12 *Possible transformer earthing arrangements*

Philosophy of protection

As well as providing protection for the usual main items of generation and distribution such as generators, transformers and transmission lines or cables, it is also necessary to decide on the significance of each component of the system to establish the consequences of a fault. When designing a system it is customary to provide alternative plant or system configuration to ensure maximum continuity of supply in the event of any fault, and the protection adopted should be selected to implement changeover to this emergency mode. Thus, if there is only one main switchboard on the system, a fault on this will shut the system down and no protection system can prevent this. However, if a system comprises two stations with generation and switchboards, it is practicable to provide zone protection on both switchboards so that a fault on one can be detected and the board isolated rapidly from the system, leaving the other station to supply the residual system. Such rapid protective systems are available that can result in transient stability for the system, provided that adequate generation is available to support the residual load or, if not, to initiate a 'crash' load-shed system to disconnect the least important loads and leave the existing generation capable of supporting the resultant essential load.

This same logic can be applied to all sections of the network, and can include transmission lines and cables, to identify the location of any major fault and isolate it.

The transient stability study will have determined what zones require such protection and which can be dealt with as integral units.

As already indicated, the behaviour of any single parameter in the system is not sufficient to identify the significance of any single fault or determine the optimum control actions most likely to benefit the recovery of the system. This is particularly true for small systems. The behaviour of more than one parameter is necessary, together with the values before the fault and also following clearance of the fault.

This is principally due to the electromechanical behaviour of the generator sets in conjunction with that of the electrodynamic loads on the system. During any fault there will be a complicated interchange of stored energy between the generation sets and the loads and this will be affected by what form the loads take and how they interact with the process system. In many instances they can function as electrical generators during transient conditions and hence affect the system frequency behaviour.

An electrical system fault reduces the load on the generators, which will overspeed and raise the system frequency while the voltage is reduced, until the fault is isolated. If the fault is in the generator zone, the zone protection should isolate the generator from the busbars and simultaneously reduce fuel to the prime mover to minimize the overspeed. Meanwhile the rest of the system will experience a voltage recovery resulting in increase of load demand on the generators, with a consequent reduction in system frequency. To optimize the governor action, the fault conditions of frequency increase and voltage reduction should be ignored for the fault clearance time, when the two parameters will be reversed and governor action to increase fuel should be expedited.

The loss of the faulty generator will result in a nett power deficiency on the system unless it was operating with full set redundancy. This condition should initiate load shedding such that, when normal system frequency is restored, the MW demand of the system will be within the capacity of the remaining generators.

Should the system fault occur on a switchboard busbar, the whole switchboard (or disconnectable section) needs to be isolated. Hence all its associated generators will be lost to the system, together with all loads fed from it. If it had previously been in an importing mode, the remaining system should be stable, but if it had been exporting power then load shedding would be required to enable the remaining system to recover.

A fault in a transmission zone will leave all generation and loads connected to the switchboards and recovery of the system will depend upon the relative generation capacity and loads of the segregated sections of the system. When there is an alternative transmission link, the impedance of this and its load-carrying capacity will determine whether system stability can be maintained, and if so what changes in tap-changer position will be required.

Processing the data to determine these actions requires devices more elaborate than simple relays. Sophisticated relays now available are capable of making the necessary logic decisions. However, on small systems it is convenient to utilize a single processor to integrate the necessary data and determine the necessary actions: this can be incorporated into a power management system (PMS).

Reclosure

Another group of activities which can be initiated by the protection system, either with or without the assistance of the PMS, is associated with isolation of a subsystem from a larger 'main' system (Figure A2.13). When two systems, each with generation plant, become disconnected it is theoretically possible for them to continue to operate in synchronism at exactly the same frequency, voltage and phase angle, and in this condition they could be reconnected, without disturbance, to form a single system again.

However, the probability is that this condition would not apply. The generated voltages would differ in phase voltage and it would be dangerous to try to reclose the circuit breaker or switch which was opened to separate the subsystem. In the extreme case, the voltages could be of equal magnitude but in phase apposition across the switch; closing this would be more severe than applying a full three-phase short-circuit on the two systems, with the possibility of

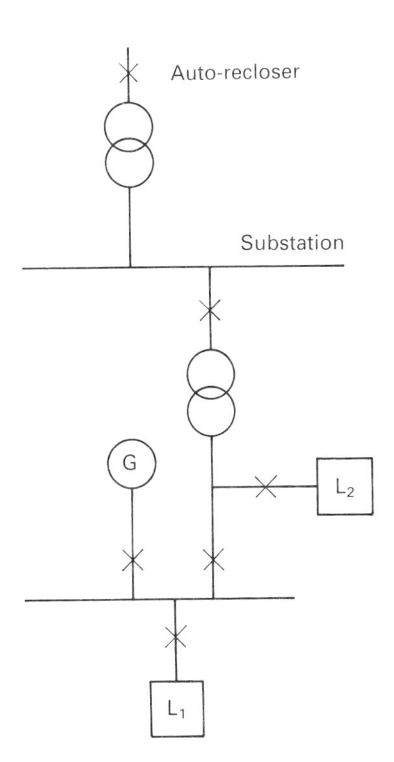

Figure A2.13 *Local generation connected through non-dedicated feeder*

causing severe damage to the generators and prime movers, and possibly to any power transformers between the generators and point of reclosure.

Such auto-reclosure of an isolating switch can be deliberately provided for by the installation of an auto-reclosing line switch on a feeder which supplies plant capable of operating as a generator. Alternatively it can occur accidentally when an operator tries to rectify an error by reclosing a switch which had been opened incorrectly. It is usual to fit check synchronizing relays on generator circuit breakers or bus section switches where accidental reclosure could be expected, but it is impractical to fit such devices on every switch on a system that could produce this dangerous condition. It is possible to fit a voltage lock-out relay on auto-reclosing breakers at any location. This will prevent auto-reclosing when it is dangerous to do so, but have no effect under healthy reclosing conditions. Alternatively, interlocking can be provided to ensure that the auto-reclosing breaker opening will trip a circuit breaker, which will isolate the generation downstream before reclosure can occur.

When it is impractical to prevent possible auto-reclosure, it is necessary to ensure that plant installed is able to cope with the effects of a full short circuit. Specially reinforced machine and transformer windings can be used, but these involve extra cost and may have other adverse effects; torque-limiting couplings can be provided, which will prevent the full magnitude of the electrically generated short circuit torque being transmitted to the gearbox or prime mover on a generator set.

Current limiters provide another means of preventing damage to electrical equipment and associated mechanical plant from the effects of large short circuit currents generated by reclosure. These devices detect the rate of increase of current and can operate before the current can reach its prospective level, in the same way as HRC fuses operate on feeder circuits.

Isolation detection

Protection against reclosure of supply can be eliminated if isolation of the downstream generation system can be positively identified and action taken to trip a non-auto-reclosing breaker, leaving the local system isolated and operable. Synchronizing features are usually provided on such a breaker so that the local system can be safely reconnected to the main system when conditions have been returned to normal.

Such detection, however, is not always a simple operation. It depends on the configuration of the main system, the method of interconnection and any associated voltage control, and the configuration and control procedures provided for the local system.

When the local system always has a nett input of power from the main system, should isolation occur the frequency of the local system will start to reduce and, if no other loads are supplied by the interconnector (i.e. it is a dedicated feeder), the detection of no power flow is a positive indication of isolation. Frequency may not be a positive indication, as governor action could be arranged to restore this to normal if sufficient load shed were initiated to leave the system stable and viable.

Where the feeder is not dedicated to the local system, when isolation occurs the local generation may supply other loads fed from the same feeder and no-power detection is no longer available as an indication of separation. Likewise, it is not valid should the local system ever operate normally to provide power to the main system. An alternative solution in the case of a dedicated feeder is to measure the power provided to the local load and compare this with the actual generated local power output. If these are equal, no import/export is occurring but this does not necessarily prove isolation. It is thus necessary to arrange the local generation governor control to ensure that this value is never zero under normal conditions; then, when it registers zero this indicates isolation. When the interconnector is not dedicated to the local system it is necessary to measure the input to the other loads fed from the interconnector and include this in the above power comparison.

This same logic can be applied to MVAR generated locally and MVAR absorbed locally, arranging constant MVAR generator excitation control to ensure that this is never normally zero, and hence identifying isolation when zero MVAR is measured at the interconnector incomer.

Ideally the combination of MW and MVAR measurement and control can detect isolation rapidly and accurately, but these factors must be considered carefully when determining the optimum operating mode for the local system to ensure complete compatibility.

Fast reclosure

Auto-reclosing breakers are in use with reclosing times as short as one second and in such installations very fast detection of isolation is required.

To get such a response is not always practicable. Some form of lock-out of the reclosure must be provided or alternatively the breaker itself must initiate the action signal. It is possible to obtain an improved response from change in system frequency, which would be the normal consequence of isolation of an island system. Such 'rate of change of frequency' relays are now in common use for detecting such conditions, where the island system parameters are suitable.

Appendix 3
Electrical system analyses

When considering island or satellite electrical generation systems it is necessary to carry out a study of the prospective system to ensure that the necessary operating criteria will be met during both normal and abnormal conditions and that all plant will remain within its designed operating range. This study should also include the effects of system growth or changes which are to be expected.

Small systems involving only a few generators can be studied by a simple manual processes with reasonable accuracy for all practical purposes, but should the results be marginal regarding the suitability of the plant, equipment or control provided, either the marginal equipment should be modified to indicate an acceptable safe operating margin, or a more rigorous study should be performed.

On larger systems, or where complexity results in significant interaction between various major components, as in generating stations, a simple study will not be capable of determining accurately such interaction. A full comprehensive study will be required if satisfactory operation is to be guaranteed.

There are three significant factors involved in all such studies:

1 The use of a standardized system of units for presentation of the necessary input data (a per unit system).
2 The use of an accurate simulation of all components and functions involved in the system (modelling or functional representation).
3 The validation of the combination of (1) and (2) used to represent the system over the complete range of parameters required for all

steady-state and transient conditions likely to be encountered by the system.

Units

Over the past few years a large variety of units have been used and superseded in the physical sciences; unfortunately, some less well-known units keep cropping up and throwing existing calculation procedures into chaos. Conversion tables are now fairly extensive but their use can still lead to inadvertent errors. While this basic problem will exist for some time yet, the situation has expedited the use of per-unit systems, not only for electrical equipment but also for associated equipment.

Electrical equipment has a normal value of rated voltage and current but has other parameters such as resistance and reactance expressed in ohms. A simplification of expression was developed by deriving the voltage drop in the impedance at rated current and expressing this as a percentage of rated voltage (IZ/V). Thus a 10 per cent impedance gave an immediate impression of the relative magnitude of the machine impedance. This system has the inconvenience that to evaluate the impedance voltage drop at half current (50 per cent), the product of current (50 per cent) and impedance (10 per cent) resulted in a value of 500, which had to be divided by 100 to give the correct answer of 5 per cent voltage drop.

By simply replacing the percentage by an expression 'per unit', and still using rated current as unity, and rated voltage as unity, the reactance voltage drop became 0.1, and at half current 0.05, thus simplifying the arithmetic.

However, the concept of a per-unit system has developed from this simple example to provide a tool which is virtually essential when dealing with more complex problems and electrical systems. It has the added advantage that in such a system the values of parameters obtained are the same irrespective of the absolute system of units in which the values were originally expressed.

Voltage regulation

This function illustrates a simple use of the per unit system for a single generator.

Consider the simple example of the rise in voltage obtained when load is thrown off an AC generator running with constant excitation and speed. The regulation is defined as $E - V/V$, where V = machine terminal voltage per phase and E = machine excitation voltage per phase.

$$\text{Then} \quad V = V \cos \phi + jV \sin \phi \text{ (see Figure A3.1)}$$

$$E = (V \cos \phi + IR) + j(V \sin \phi + IX)$$

$$\text{Regulation} = \frac{E - V}{V}$$

$$= \frac{\sqrt{[(V \cos \phi + IR)^2 + (V \sin \phi + IX)^2]} - V}{V}$$

where
R = machine resistance per phase
X = machine reactance per phase
I = machine current per phase
ϕ = machine power factor

Now if V is the rated terminal voltage and I is rated current, then the resistance R and reactance X ohms expressed in per unit are:

$$\frac{IR}{V} \text{ and } \frac{IX}{V}$$

Therefore, the regulation is equal to:

$$\sqrt{\left[\left(\cos \phi + \frac{IR}{V}\right)^2 + \left(\sin \phi + \frac{IX}{V}\right)^2\right]} - 1$$

$$= \sqrt{[(\cos \phi + R/\text{p.u.})^2 + (\sin \phi + X/\text{p.u.})^2]}$$

and the regulation at any other current i amperes, or i/I per unit is:

$$\sqrt{(\cos \phi + i/\text{p.u. } R/\text{p.u.})^2 + (\sin \phi + i/\text{p.u. } X/\text{p.u.})^2]} - 1$$

where $i/\text{p.u.} = i/I$, $R/\text{p.u.} = IR/V$, $X/\text{p.u.} + IX/V$

All the values in these expressions are dimensionless ratios so this equation is independent of actual values of amperes, volts and ohms, i.e. independent of size and rating of the machine. The results

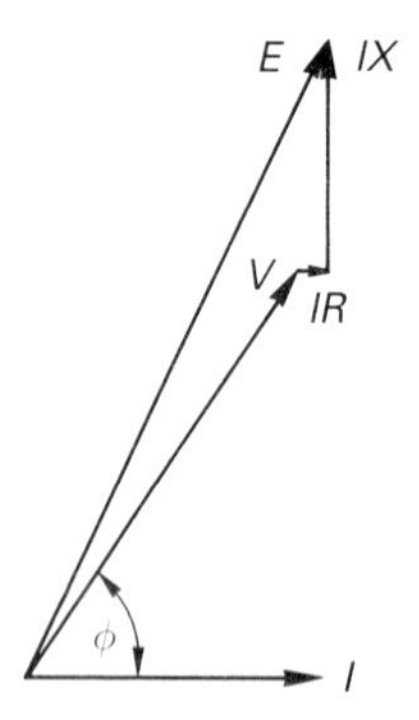

Figure A3.1 *Simple vector diagram for AC generator*

are, in fact, the same for all machines having the same values of R p.u. and X p.u.

This is one great advantage of per unit values, namely, that generalized data such as equations and curves can be derived and used for a complete class of machine irrespective of size and rating. This point will become even more obvious after considering other applications of the system.

Synchronous machine reactance

Other functions can be included in the simple per unit system, e.g. the AC generator excitation in the following analysis.

Figure A3.2 shows the combined voltage and mmf (magnetomotive force) vector diagrams for a synchronous generator. They can be combined in this way because the machine rotor operating at synchronous speed maintains its space relationship with the air-gap flux and the stator winding fields. Neglecting saturation, leakage and resistance initially, the air-gap flux Φ required to produce a terminal voltage V requires field current flowing in the field winding to provide mmf AT_e. When the load current I flows at a lagging power factor cos ϕ, the stator winding exerts demagnetizing armature reaction mmf and the field winding must produce an equal and opposite amount of mmf AT_{ar}. Thus the total excitation mmf required is AT_f, the vector sum of AT_e and AT_{ar} (Figure A3.2(a)).

If we now select the unit base of mmf as the value required to produce the unit base of voltage V on open circuit, then AT_e is the unit base of mmf. To adjust the mmf diagram to this unit base it is necessary to divide all the values by AT_e and the three values AT_e,

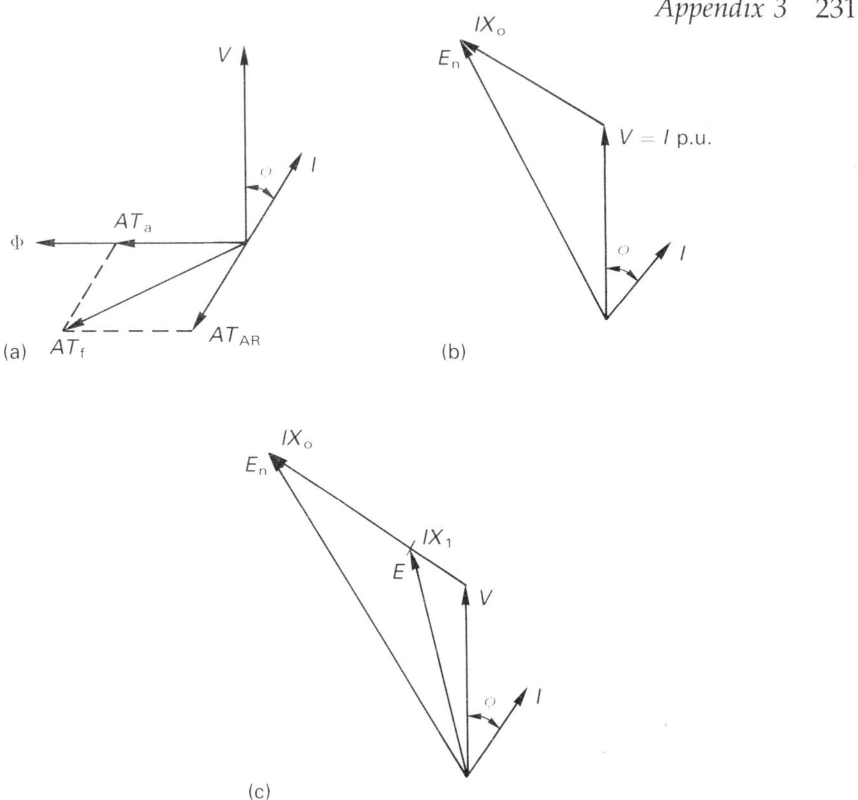

Figure A3.2 *Development of simple vector diagram*

AT_{ar} and AT_f become the ratios 1, AT_{ar}/AT_e and AT_f/AT_e, and this last ratio is usually represented by E_n. Now if the current I is in per-unit and AT_{AR} is the value of armature reaction ampere-turns corresponding to unit base current, then $AT_{ar} = I \times AT_{AR}$ and $AT_{ar}/AT_e = IAT_{AR}/AT_e$. The ratio AT_{AR}/AT_e is usually represented by the symbol X_a which is called the armature reaction reactance.

By substituting these new per-unit values in the mmf diagram one can combine it with the voltage diagram quite simply by rotating it through 90° so that the unit values of V and AT_e coincide (Figure A3.2(b)). The simple voltage diagram thus obtained is the same as that representing a circuit having a generated voltage E_n per unit and a series reactance X_a per unit and giving a terminal voltage V (which is the base unit) across a load such that the current flowing is I per unit at a power factor cos ϕ lagging.

It will be noted that E_n is the per unit terminal or open circuit voltage which would be generated by the AC generator with the same value of excitation, and that the voltage regulation is therefore $(E_n - V)/V$ or $E_n - 1$ since V is unity. This simple relation only applies when no saturation occurs, of course.

In practice, stator leakage reactance X_1 exists, and when this is included one gets the voltage diagram shown in Figure A3.2(c). The stator leakage reactance X_1 and the armature reaction reactance X_a can conveniently be combined and the sum $X_d = X_1 + X_a$ is known as the synchronous reactance. The stator resistance can also be included, and the vector sum of this and X_d is sometimes known as the synchronous impedance. When saturation exists the ratio AT_{AR}/AT_e does not remain constant for all values of voltage and current.

Mechanical systems

The per unit system can also be extended easily into associated systems, such as prime movers or driven mechanical plant.

When rotating machines are being considered, mechanical power is an important factor, either as input to a generator or output from a motor, and this value is related to the electrical power by the efficiency. The base electrical unit values considered so far can be regarded as the output of a generator or the input to a motor and the product of unit current and unit voltage can be defined as unit electrical power. This unit electrical power is based on kVA and is related to kW by the power factor, and to mechanical output by the quotient of efficiency and power factor in the case of a generator and by the product of efficiency and power factor in the case of a motor. It is obviously convenient to keep the per unit electrical impedance of similar sizes of machines equal whether they are generators or motors, or a single machine which can operate as either. This purpose can be achieved if one considers electrical per-unit power as defined above for both generators and motors.

The most important use for mechanical power is to determine conditions when speed changes occur, e.g. during transient conditions or when starting and accelerating as a motor, and it is necessary to determine the torque corresponding to the power. The torque corresponding to a power of kVA at a synchronous speed of N rev/min is:

$$\text{kVA} \times \frac{1}{0.746} \times 550 \times \frac{60}{2\pi N} = \frac{\text{kVA}}{N} \text{ ft lb}$$

and if all torques are expressed in terms of the torque corresponding to unit electrical power (kVA) then a per unit torque T p.u. corresponds to T p.u. 7040 (kVA)/N ft lb.

The acceleration of a machine having inertia I ft lb/s^2 produced by a torque T ft lb is:

$$\frac{d^2\theta}{dt^2} = \frac{T}{I} \text{ rad/s}^2$$

and if one introduces a parameter H such that:

$$I = 2H \times (550\,\text{kVA})\ \omega^2\ 0.746$$

and $\omega = (2\pi N)/60$ rad/s, then the acceleration due to per unit torque T p.u. is:

$$\frac{d^2\theta}{dt^2} = \frac{T\ \text{p.u.}}{2H} \times \omega \text{ rad/s}$$

and if ω rad/s corresponds to N rev/min, which one can regard as unit speed, then:

$$\frac{d^2\theta}{dt^2} = T\,\text{p.u.}\ 2H$$

per unit speed per second.

This simple equation is based on the new system of units, namely, unit power equal to kVA, unit speed equal to N rev/min (synchronous speed) and unit time equal to 1 s. The value H which was selected apparently arbitrarily is also an important unit value. The stored kinetic energy in the machine with inertia I at N rev/min is:

$$\frac{1}{2} = I\omega^2 \text{ ft lb}$$

$$= H\,\text{kVA}\,\frac{550}{0.746} \text{ ft lb}$$

$$= H\,\text{kVA kW s}$$

Thus

$$H = \frac{\text{stored kinetic energy (in kW s)}}{\text{unit kVA}}$$

in seconds dimensionally, and is the stored energy constant, although it is sometimes referred to erroneously as the inertia constant.

A system of units has thus been derived which is consistent with those used for electrical problems but which includes mechanical characteristics. Thus for synchronous machines under balanced operating conditions one can use as base units:

Rated phase voltage
Rated phase current
Rated speed (synchronous)

and derive unit impedance, unit power, unit torque and the stored energy constant H, assuming unit time as the second.

These form a coordinated system of units which permits the solution of problems involving both electrical and mechanical variables of a machine in a simple manner and which permits generalized solutions. Thus the solution of problems using typical parameters is possible and actual machine values are not essential for a general solution.

Figure A3.3 illustrates such a generalized solution to the problem of the run-down times of a machine with a total stored energy

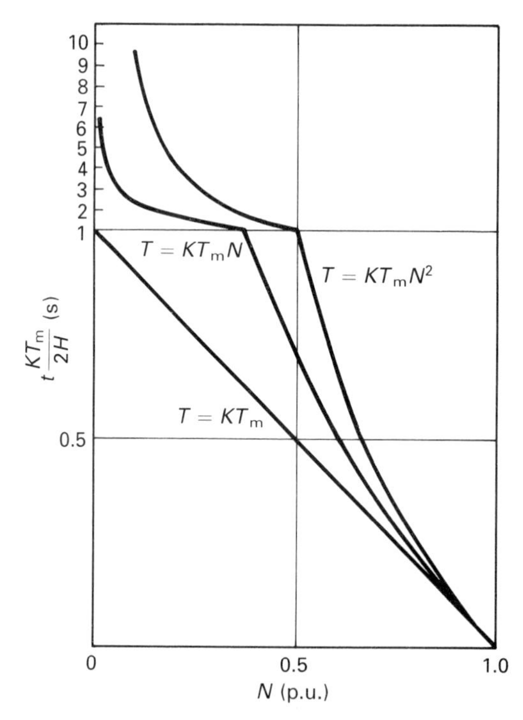

Figure A3.3 *Generalized curves for deceleration times*

constant H for three alternative forms of braking torque, namely, constant torque, torque proportional to speed and torque proportional to the square of the speed. From these curves can be deduced the time interval between any per unit speeds for any of the forms of torque applicable.

Induction motors

The per unit system developed for synchronous machines, generators or motors uses the same bases for rating, i.e. the motor input is regarded as unity. This has the advantage that the capability of regeneration and fault contribution results in no differences between motors and generators.

This same approach has been used for induction motors and consequently simplifies the analysis of combined systems, including mechanical features.

Table A3.1 Induction motor equivalent circuit values

Machine	Base values			
	V	I	kVA	Z
A	3810	741	8470	5.14
B	6350	37.5	714	16.92

	Ohms		Per unit	
	Machine A	Machine B	Machine A	Machine B
R_S	0.025	2.04	0.005	0.012
X_S	0.652	17.3	0.126	0.102
X_M	24.9	560.0	4.85	3.31
R_R	0.0324	1.03	0.006	0.006
X_R	0.60	17.15	0.117	0.101

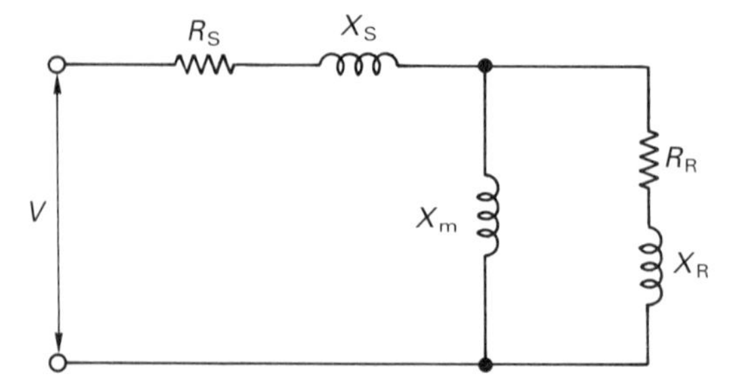

Figure A3.4 *Simple equivalent circuit for induction motors*

Table A3.1 shows the selected base unit values for two designs of induction motor, and also values for the simple equivalent circuit shown in Figure A3.4 expressed in ohms and also in per unit. This illustrates immediately the value of such per unit values, which are very similar although the actual machine ratings are very different. The relatively high per unit stator resistance of machine B occurs because it is a comparatively small machine rated at 11 000 line volts, and is quite a typical value. Its magnetizing current is somewhat higher than that of machine A but otherwise the per unit values correspond very closely.

It is most important to realize that the mathematical transformation used to derive the simple circuit shown in Figure A3.4 has eliminated the complication of slip frequency and, in fact, the powers and torques are 'synchronous', i.e. they are based on synchronous speed. Thus, to obtain the shaft torque in ft lb it is first necessary to determine the rotor I^2R in watts and then multiply by 7.04 and divide by the synchronous speed in rev/min. Thus, if one uses the same values of unit power and torque suggested for synchronous machines then per unit torque is the ratio of rotor I^2R to kVA. This per unit torque can be used in the manner developed previously for synchronous machines. A typical per unit torque/per unit speed curve for an induction motor is shown in Figure A3.5.

Starting characteristics

Motor starting characteristics are determined by a comparison of the motor torque speed characteristics with those of the mechanical driven unit, and these latter are usually expressed as a percentage of full-load torque related to percentage full speed. It is thus necessary

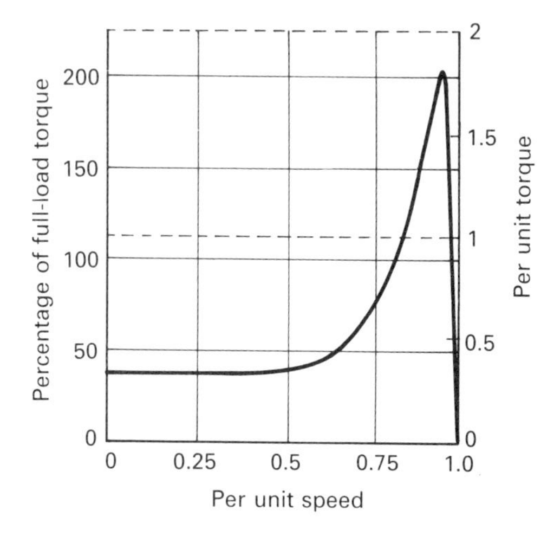

Figure A3.5 *Typical torque–speed curve for an induction motor*

to convert motor per unit torque to percentage full-load torque or convert the latter to per unit torque, and in the interests of standardization of procedure this latter operation is the better one.

To obtain the relationship between per unit torque and percentage full-load torque the following symbols are used:

HP = Rated full-load horsepower
PF = Rated full-load power factor
EFF = Rated full-load efficiency

kVA = rated full-load input (electrical)

$$\therefore \quad \text{kVA} = \frac{\text{HP} \times 746}{\text{PF} \times \text{EFF}}$$

N_{FL} = rated full-load speed
S = rated full-load slip
N = synchronous speed $\qquad \therefore N_{FL} = (1 - S)\,N$
W = rotor I^2R watts

Then output torque $T = 7.04(W)/N$ ft lb and full-load torque $T_{FL} = 5250(\text{HP})/N_{FL}$ ft lb. Therefore,

$$\text{percentage torque} = \frac{7.04 WN_{FL}}{5250 \ \text{HP} N} \times 100$$

$$= \frac{W \ (1 - S)}{\text{kVA} \times \text{PF} \times \text{EFF}} \times 100$$

and base torque = 7040 (kVA)/N ft lb

$\therefore$ p.u. T = W/kVA

$\therefore$ percentage torque/p.u. T = $(1 - S)/\text{PF} \times \text{EFF} \times 100$

For example, consider a 1000-h.p. 1000-rev/min motor with full-load efficiency of 0.95 and full-load power factor of 0.92; then the input kW is 786 and kVA is 855. If the supply is 3300 V, three-phase, full-load line current is 149.5 A, and if one uses base unit voltage of 1905, then base unit impedance is 12.75 ohms. With a full-load slip of 1 per cent, full-load speed is 990 rev/min and full-load torque is 5300 ft lb.

Now, base unit torque is 6019 ft lb and

$$\frac{1 - 0.01}{0.92 \times 0.95} \times 100 = 113.3$$

and therefore 100 per cent full-load torque corresponds to 100/113.3 = 0.88 p.u. T and 5300/6019 = 0.88 p.u. T, which is the correct value.

When the torque–speed curves of both the motor and its driven load are available it is possible to check that starting will be satisfactory and also to determine the starting time and derive the starting current–time curves. This can be done fairly easily when the inertias of the motor and load are known and all the various units are adjusted to the correct values. However, here again the use of wrong or inconsistent units can lead to errors and the calculations may require to be checked, thus wasting time. For example, the inertia of a load may be described in the forms WR^2, WR^2/g, GD^2 and in units such as tons ft^2, lb in^2, kg m^2 etc., the form of the value is not always clearly stated and errors often occur when converting to the usual system of units. A very simple calculation is possible if per unit values are used. It can easily be shown that the time required to accelerate a system having a stored energy constant H per unit over an interval of per unit speed, k per unit, when a

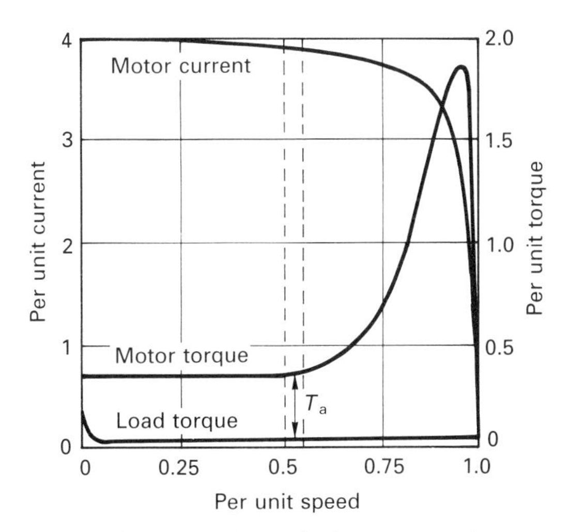

Figure A3.6 *Starting calculations, step by step*

constant torque of T_a per unit is applied is $2Hk/T_a$ s. Thus it is only necessary to plot the motor and load per unit torque/per unit speed curves on the same graph, divide the base into a convenient $(1/k)$ number of equal intervals and evaluate the average difference or accelerating torque T_a over each interval (Figure A3.6). If the intervals are suitably chosen, this value can be evaluated very easily by inspection. If the motor line current is plotted to the same per unit speed base the values of this at the beginning and end of each speed interval can be included. The plotted results of such a

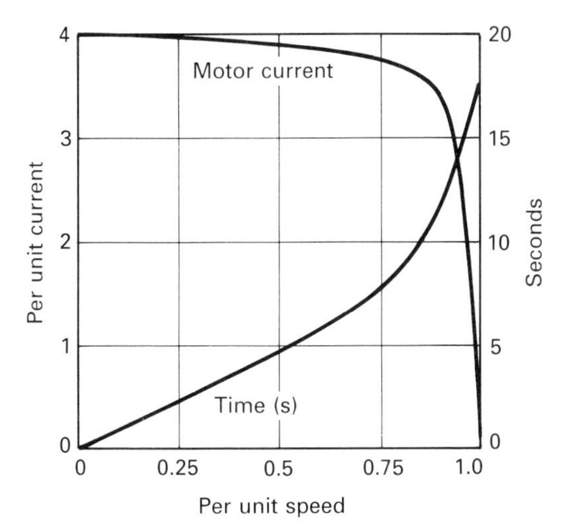

Figure A3.7 *Starting current–time characteristic*

calculation are shown in Figure A3.7. Over most of the torque curves it will be found that the difference torque at the middle of the interval is equal to the average accelerating torque, and in many instances this value only need be measured and the calculation then becomes a very simple and rapid one. Many load characteristics are typical, and for a given type of motor characteristic typical curves of current–time can easily be derived for a range of H values. The amount of labour saved is obviously very great for each individual machine and each load characteristic. Typical values of T_a and H are known, and the simplicity of having no 'units' to worry about ensures that the possibility of calculation mistakes is virtually zero.

Machine natural frequency

Another commonly encountered machine problem involving mechanical data is associated with 'hunting' or 'swinging', resulting in torsional oscillation of the machine rotor system. The synchronous torque of a synchronous machine is a function of the displacement or load angle by which the rotor mmf lags or leads the stator flux, and if the rotor oscillates about the steady synchronous speed the torque will also fluctuate. The rotor possesses inertia and it is obvious that if some factor can provide an exciting disturbance of a particular constant frequency, then the machine can act in an analogous way to a mass suspended by a spring, and sustained oscillations may occur. At a particular critical or resonant frequency the machine could, theoretically, be forced out of synchronism. In practice appreciable damping is always provided, and some machines have been built which actually operate satisfactorily while running at this critical speed because of the large amount of damping which they possess.

The calculation of the natural frequency of oscillation of a synchronous machine assumed to be connected to an infinite supply, i.e. the terminal voltage is assumed to remain constant throughout the period of oscillation, can be determined as follows.

The natural frequency of a rotor with a moment of inertia I ft lb/s^2 and a torque of amplitude T lb ft is

$$f_o = \frac{1}{2\pi}\sqrt{\frac{T}{I}} \text{ per second}$$

If the electrical synchronizing power is assumed to be proportional to displacement:

$$f_o = P_o \times \text{kVA kilowatts per electrical radian displacement}$$

$$f_o = 7040\ P_o\ \text{kVA}\ (P)/2\ \text{lb ft per mechanical radian,}$$

where P = number of poles.

Substituting $120f/N$ for P and WR^2/g for I,

$$f_o = 35\,208 - \sqrt{\left[\frac{P_o \times \text{kVA} \times f}{N^2 WR^2}\right]} \text{ per minute}$$

However, if one uses the coordinated per unit system as already developed, the corresponding equation is

$$f_o = \frac{30}{\sqrt{\pi}} \sqrt{\left[P_o \times \frac{f}{H}\right]} \text{ per minute}$$

$$= 16.93 \sqrt{\left[P_o \times \frac{f}{H}\right]} \text{ per minute}$$

$$= 16.93 \sqrt{f} \sqrt{\frac{P_o}{H}} \text{ per minute}$$

and as typical values of P_o and H are known for specific types of machine, a typical answer is quickly obtainable for any class of machine.

From this equation a general set of curves can be drawn, giving values of $f_o \sqrt{(50/f)}$ for useful ranges of P_o and H (Figure A3.8). Simple curves can be drawn for the value of P_o in terms of X_d and X_q per unit, and for each type of machine typical values of these are known and hence typical values of P_o. It is consequently a simple matter to select a value of H to give a natural frequency remote from any known forced frequency.

Machines operating in parallel

The effect of sets operating in parallel is to vary the voltage at the machine terminals and consequently vary the value of P_o on each machine; that is, P_o, instead of being constant, will fluctuate depending on the oscillations of any other machines on the system.

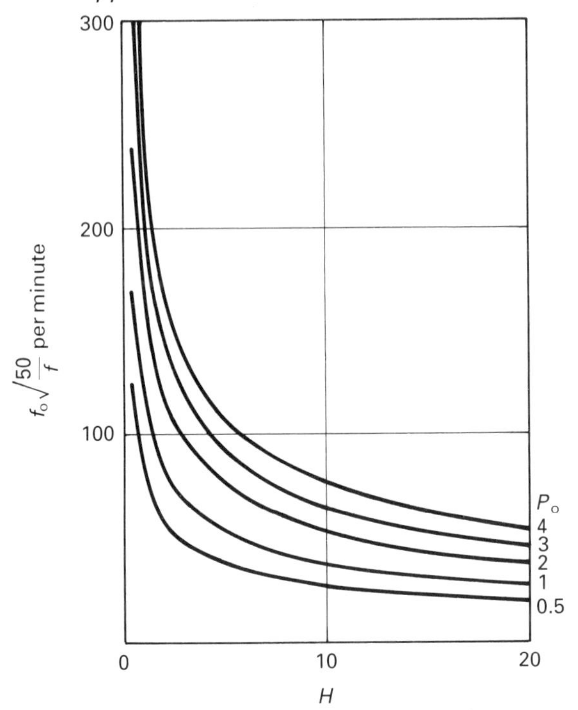

Figure A3.8 *Values of* $f_o\sqrt{(50/f)}$ *for useful ranges of* P_o *and* H

If two machines are fed from an infinite supply they have their own natural frequency but, as the relative amount of impedance between the machines and the source of constant voltage increases, the voltage fluctuation will increase until the condition is reached when the two machines form an isolated two-machine system. The question of natural frequency of machines in this condition has been investigated, and these two limiting conditions give some guide to possible intermediate conditions. Various approximate formulae have been produced to give a combined natural frequency of two or more machines but they do not give accurate results for all conditions. These can only be derived by a full system analysis or tensor study, which is obviously most easily carried out on a per unit basis.

Machine transient stability

Where the behaviour of a specific drive on a system is being assessed it may be undesirable to undertake a full stability study. It is usually possible to assess the behaviour of a single unit fairly accurately using typical per unit values. It is then possible to adjust these values to obtain a more desirable performance and then use these as a basis for obtaining a suitable unit. When this stage has been reached, an accurate study may be justified to confirm the approximate analysis already carried out. This procedure is easily performed on large synchronous drives which should be selected to have optimum characteristics to benefit the system.

During transient stability studies it is usual to assume that a synchronous machine generates its transient torque. This ignores subtransient effects and the generator set decrement but it appears in practice that these and other secondary factors tend to compensate for each other to justify this basic assumption.

In practice, for transient conditions the flux is represented by a fictitious 'voltage behind the transient reactance'. E' and the transient power is a function of the quotient of this voltage and the transient reactance. The actual value of transient reactance which should be used is that related to rated voltage conditions. It is quite practicable to use this quotient, known as the 'transient parameter', as a single parameter to determine the transient power characteristic of a synchronous motor.

Transient stability of synchronous machines has usually been determined by a method based on the 'equal-area criterion', and experience over past years has shown that this assumption is safe.

Generators usually possess relatively low damping and it is quite customary to ignore the effect of this on stability, certainly in an approximate investigation. Motors invariably have an appreciably higher degree of damping than corresponding sizes of generators. Consequently, results for motors are even more pessimistic than for generators and so help to make more realistic the results of a generalized approach to the problem for motors.

Thus, assuming load torque constant at 1 p.u. and evaluating stability on a time base of $t \times H^{-1/2}$, it is possible to determine the value of motor terminal voltage which will result in transient stability for a machine parameter based on E'/X'_d, the transient parameter.

A typical set of curves calculated using the methods mentioned previously is given in Figure A3.9. These assume that the recovery voltage is adequate to give static stability.

From these curves it can be seen that with a value of H equal to unity, which is low for most applications, a fault clearance time of

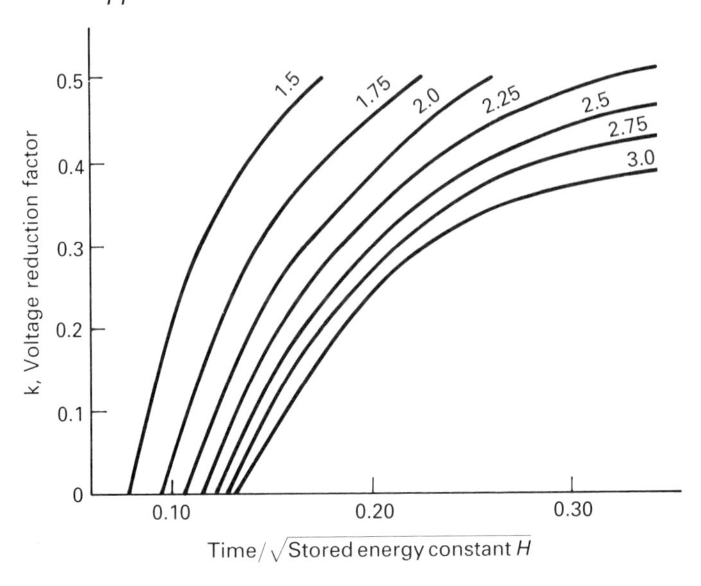

Figure A3.9 *Curves relating reduced voltage and time function for various values of transient parameter* E'/X_d'

0.2 s can result in stability for a drop in terminal voltage to 0.35 p.u. if the machine transient parameter is 2.25. If the system condition required a lower value of k, or a longer time t, then this could be met by adjusting either H or the transient parameter. The cost increases associated with these alternative means of obtaining stability are quite different from each other and also vary depending on the type of machine. Since the general type of motor is known in any particular instance, however, it is a relatively simple matter to determine what measure of adjustment of these two machine parameters simultaneously can result in meeting the required stability condition most economically.

The most limiting condition is $k = 0$, i.e. zero voltage during the disturbance. It is important to note that even with $k = 0.1$ the limiting clearance time for stability can be increased by 15 per cent, and for $k = 0.2$ the increase can be 36 per cent over that for $k = 0$. It is essential in the interests of economy to determine the actual value of minimum voltage during a disturbance rather than to assume, for simplicity, that it is zero.

Figure A3.10 shows the relationship between the machine parameters and the time for $k = 0$, and indicates that the rate of increase of t is greater when H is increased than when the transient

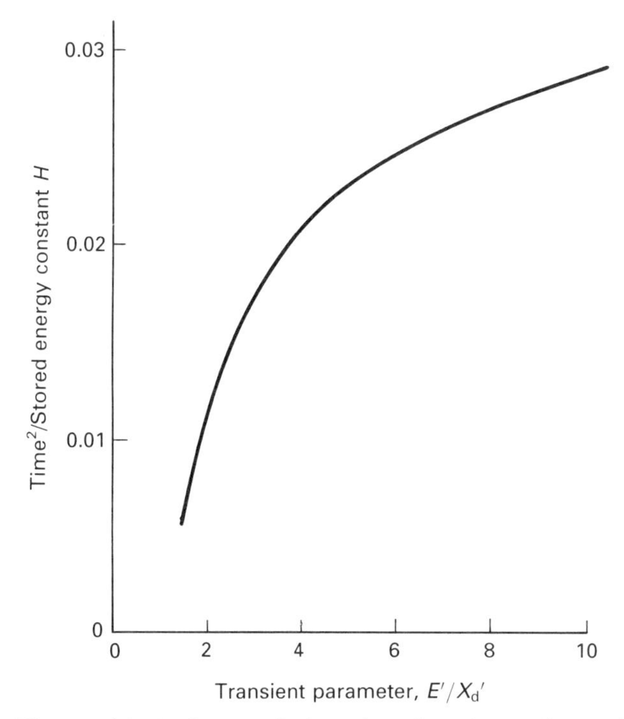

Figure A3.10 *Curve relating time function and transient parameter for zero voltage*

parameter is increased. To a first approximation the value of H is proportional to the product of the transient parameter and the square of the machine diameter. It is therefore usually more convenient and practicable to increase the machine diameter and reduce its length to increase the value of H while maintaining the transient parameter constant, than to reduce the diameter and increase the length to increase the transient parameter while maintaining a constant value of H.

In practice it is usually preferable to increase the H value of the machine to obtain a higher value of t. When the amount of increase required is considerable, however, a simultaneous increase of transient parameter is usually found to be desirable. To obtain a large increase in time, t, the transient parameter will have to be effectively increased in conjunction with an increase in the value of H.

As well as varying the economic values of inertia and transient characteristics inherent in a machine, it may be possible to modify the characteristics of equipment outside the motor. For example, on

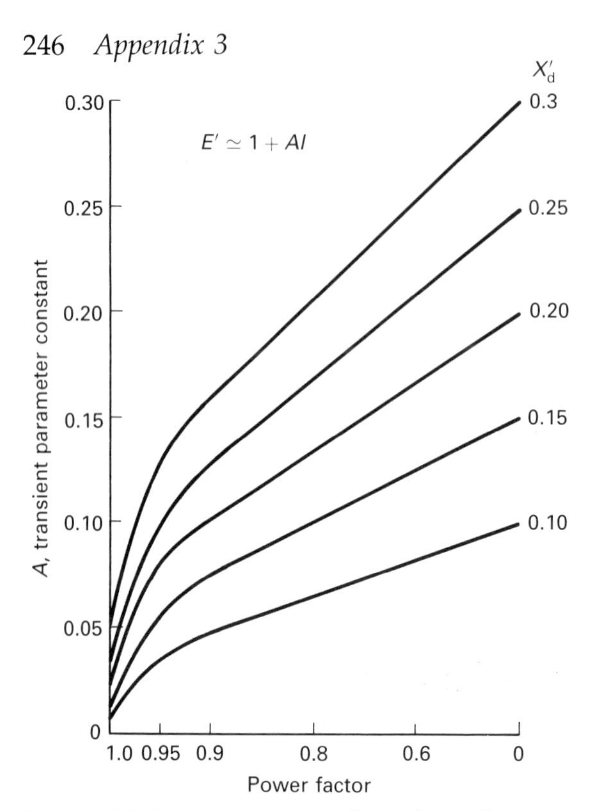

Figure A3.11 *Approximate values of transient parameter constant*

low-speed motors driving reciprocating compressors it may be more economical to add a flywheel than to increase only the motor inertia to obtain the required increase in time, t.

The transient parameter of a synchronous machine is a function of the excitation conditions and the transient reactance, and a very approximate value can be obtained using the data given in Figure A3.11. It will be noted that this value is dependent on the machine power factor and current at the instant of system disturbance. E' is the voltage behind the transient reactance.

It is still possible to retain transient stability with fairly long clearance times if a high value of transient parameter can be built into the machine. One requirement of such a design, of course, is a high value of excitation with consequent difficulties caused by saturation. When it is desirable to obtain transient stability for low values of terminal voltage and long clearance times it may be much more economical not to use the fixed excitation system on which the above analysis has been made but to use a high-response excitation control scheme.

With such an arrangement, it is necessary to compare the cost of a machine to perform satisfactorily when using fixed excitation with the combined cost of a cheaper machine and the excitation system required to give the required degree of stability. The present approach to the fixed excitation machine can be used to obtain an approximation to the requirements of a suitable forced excitation system and so simplify the investigation required to make such a comparison.

The process of recovery stability of an induction motor is quite different from that of transient stability of a synchronous motor and, because of the longer times usually involved in the former problem, it has been customary to use the 'steady-state' characteristics of induction motors for such investigations.

It is unfortunate that for ease of calculation the symbols used for induction motor analysis are sometimes the same as those already developed and standardized for synchronous machines, even though they do not represent the same conditions of operation. The use of the more correct transient state analysis for induction motors is not sufficiently developed to be adopted in a generalized form and the following is based on the conventional steady-state values only.

Single-cage motors have a well-known shape of torque–speed characteristics, and if the circuit components possess constant parameters it is a simple matter to generalize a torque–speed characteristic. This would relate the peak torque, the breakaway torque and the full-load slip, or associated parameters.

In practice, however, it is necessary only to relate the actual motor torque–speed to that demanded by the load during starting and when this is the same as that experienced during the recovery conditions, which is usually so, the only significant factors are the amount of voltage reduction on the motor (with its consequent reduction of the steady-state torque), and the time (with its consequent reduction in speed).

For the majority of induction motors the critical point is the torque available in the region between 80 per cent and 90 per cent of full speed. Above this value the torque rises to its pull-out value and it is reasonable to assume, with a load requiring rated torque, that the behaviour at, say, 85 per cent speed will determine behaviour at a higher speed adequately for the present purpose.

Thus if we can calculate the time to reach this 'critical' speed, it is possible to determine whether the motor will continue to decelerate or will start to accelerate having recovered stability. Recovery, of course, is also dependent on the recovery voltage at the machine terminals after the fault clearance. This determines the steady-state stability, in conjunction with other conditions, such as speed then prevailing, and must also be checked to ensure complete recovery.

Generalized deceleration curves are available and from these it is possible to determine the time required to change speed for a known decelerating torque. These are expressed in per-unit values, and by using a suitable per-unit system it is possible to deal with induction and synchronous motors on a common basis.

Using the most pessimistic load torque, and assuming zero terminal voltage, Table A3.2 gives times to reach 85 per cent full speed for values of $(H/T)^{1/2}$.

Table A3.2 Times taken to reach 85% full speed and to slip one pole

$\left(\frac{H}{T}\right)^{1/2}$	0.5	0.6	0.67	1	2	5	10
Time (s) to reach 85 % *N*	0.075	0.108	0.133	0.3	1.2	7.5	30.0
Time (s) to slip one pole	0.1	0.12	0.133	0.2	0.4	1.0	2.0

From Table A3.2 it will be seen that for full-load torque, *T*, and a value of *H* of 1, the time to reach 85 per cent speed is 0.3 s, which is relatively long for a good fault clearance time. For a typical pump characteristic this time would, of course, be longer, and the appropriate value can easily be obtained from the generalized curve.

Very-high-efficiency motors may have their critical speed closer to 90 per cent than 75 per cent speed, and for such the times required would, of course, be proportionately shorter than those given in Table A3.2. As this feature is now commonly required for large and important motors, this fact must be carefully considered when checking recovery stability.

The second set of times given in Table A3.2 has been included because, while rapid clearance times are desirable for the recovery stability of induction motors, the reduction in speed during disturbance is also associated with a change in phase angle between the motor voltage generated by its own flux and rotation and that of the supply voltage. Sudden restoration of the supply voltage can therefore result in a condition which is appreciably more severe than a sudden short circuit at the motor terminals. The time taken to slip one pole pitch, i.e. for the motor and supply voltages to be anti-phase, has been included in Table A3.2 as it is a function of the same general parameter. The two times are equal at a value of $H = 4/9T$, a condition not normally encountered in practice. It is important, however, to check that the clearance time involved in a particular motor problem is not such as to involve an added hazard from the effects of restoration of the supply with the voltages anti-phase. This condition also applies to synchronous motors and generators.

When it is required to determine approximately whether an AC motor is likely to be stable during a specific system disturbance and sufficient data are not available to enable a complete system study to be carried out, it is possible to utilize a generalized approach to the machine behaviour to obtain an answer.

This approach can be used to determine the limiting machine in a multi-machine system and so reduce the effort involved by requiring a detailed analysis of this single machine to determine whether the complete system is likely to be stable.

The general parameters indicate the characteristics required of a motor if it is to be stable under specific fault conditions and enable assessments to be made of the optimum or most economic means to obtain satisfactory stability, involving the machine, the system and the protection fault clearance time. Alternatively the method enables the stability capabilities of specific designs of motor to be compared without the need for an elaborate system study.

Machine system analysis

The use of per-unit values for machine analysis has been developed from the simple idea of per unit voltage and current to per unit impedance. The use of per unit mmf enables per unit machine reactance to be derived. The use of a per unit stored energy constant then enables the mechanical characteristics of the machine to be related to the existing per unit system. This coordinated system of per units is now generally recognized and is in use in machine design and in system analysis. This V, I, H, t system has proved itself very useful, particularly for system studies where V and I in dimensionless per-unit values combine easily with H per unit (in seconds) and the time per unit in seconds. The simple relation between per-unit torque, H and t, namely

$$t = \frac{2H}{T} \text{ p.u.}$$

is very convenient and is the basis of many studies of transient problems involving machines. These values, one should note, only apply to the machine viewed from its terminals and do not indicate any internal conditions such as rotor speed or currents or angular position.

In machine stability studies it is necessary to consider the machine electrical displacement angles and the equation of motion already derived, i.e.

$$\frac{d^2\theta}{dt^2} = \frac{T \text{ p.u.}}{2H} \omega \text{ mechanical radians/s}^2$$

can be modified using the relationship that θ mechanical radians correspond to δ electrical radians where $\theta = (2\delta)/P$.

Thus

$$\frac{d^2\delta}{dt^2} = \frac{2}{P}\frac{d^2\theta}{dt^2} = \frac{T \text{ p.u.}}{2H}\frac{2\omega}{P}$$

and since $P\omega = 4\pi f$, where f is the system frequency in cycles per second,

$$\frac{dt\delta}{\delta t^2} = \frac{\pi f}{H} T \text{ p.u.}$$

$$\text{or} \quad T \text{ p.u.} = \frac{H}{\pi f}\frac{d^2\delta}{dt^2}$$

which is the equation normally used.

Now it is possible, using a different system of units, to convert this equation to the more fundamental one:

$$T \text{ p.u.} = L \frac{d^2\delta}{dt_1^2}$$

where T p.u. and δ are as defined above, L is a new inertia constant and t_1 is a new time unit.

It should be remembered that the selection of values has not been made through necessity but has merely been directed by current requirements. The value H was derived from a system which has already assumed the unit of time of one second, which for the mechanical systems from which it was derived was a fairly obvious choice.

When fundamental machine analysis is being carried out, this choice of system is not found to be so suitable because the values for H and t are of mechanical derivation and are associated with the space angular rotational speed of the machine rotor. If one considers the magnetic field rotating in a synchronous machine, the system frequency determines the rate of change of electrical angle, which is the factor of interest when considering the magnetic field; this value is common for all machines connected to the same power system. The choice of per-unit time can thus be conveniently taken as the time taken for the synchronous field to rotate through an electrical radian, which is $1/\omega_1$ seconds where $\omega_1 = 2\pi \times$ supply frequency. This new time unit applies universally for all machines connected to the same supply system.

Now consider the natural frequency as already derived:

$$f_o = \frac{30}{\sqrt{\pi}} \sqrt{\left(\frac{P_o f}{H}\right)} \text{ per minute}$$

$$= f \sqrt{\left(\frac{P_o}{2\omega_1 H}\right)} \text{ per second, where } \omega_1 = 2\pi f$$

Now H, the stored energy constant based on rated kVA, is dimensionally in second units; in the new time units, which are $1/\omega_1$ seconds, the corresponding value is $\omega_1 H$ in time units; and the inertia constant, which is twice the stored energy constant, is $2\omega_1 H = L$ in the new system of units. The values H and L are both in general use and it is obvious that their correct descriptions, stored energy constant and inertia constant respectively should always be used to avoid confusion. In the new system, frequency is f and the natural frequency is now

$$\sqrt{\left(\frac{P_o}{L}\right)}$$

per unit, which is designated as h, and this fundamental equation is now

$$h = \sqrt{\left(\frac{P_o}{L}\right)}$$

In this new system of units the angular acceleration

$$\frac{d^2\delta}{dt_1^2} = \frac{1}{\omega_1^2} \times \frac{d^2\delta}{dt^2}$$

and the equation referred to previously,

$$T = \frac{H}{\pi f} \times \frac{d^2\delta}{dt^2}$$

becomes

$$\mathrm{T} = \frac{2H}{2\pi f}\,\omega_1^2\,\frac{d^2\delta}{dt_1^2}$$

$$= L\,\frac{d^2\delta}{dt_1^2}$$

which is again of fundamental form in the new system of units.

This system of units, per-unit volts and amperes, per-unit time $1/\omega_1$ seconds, and inertia constant L, will be found in most modern books and articles on machine analysis. Because of the elimination of the effects of actual rotor speeds, it permits single tensors to be used to represent machines connected to common systems.

Simulation

When the simple manual investigation of a system is found to be inadequate it is necessary to go to a comprehensive study, and the amount of data and modelling required, together with the complexity of the calculations, justifies the adoption of a computer or processor to perform this function. In the past, analog models were often used for this purpose because they allowed the operator to assess the effects of changing any parameter individually and proved a very good educational tool. However, today digital analysis is virtually universally used for system studies.

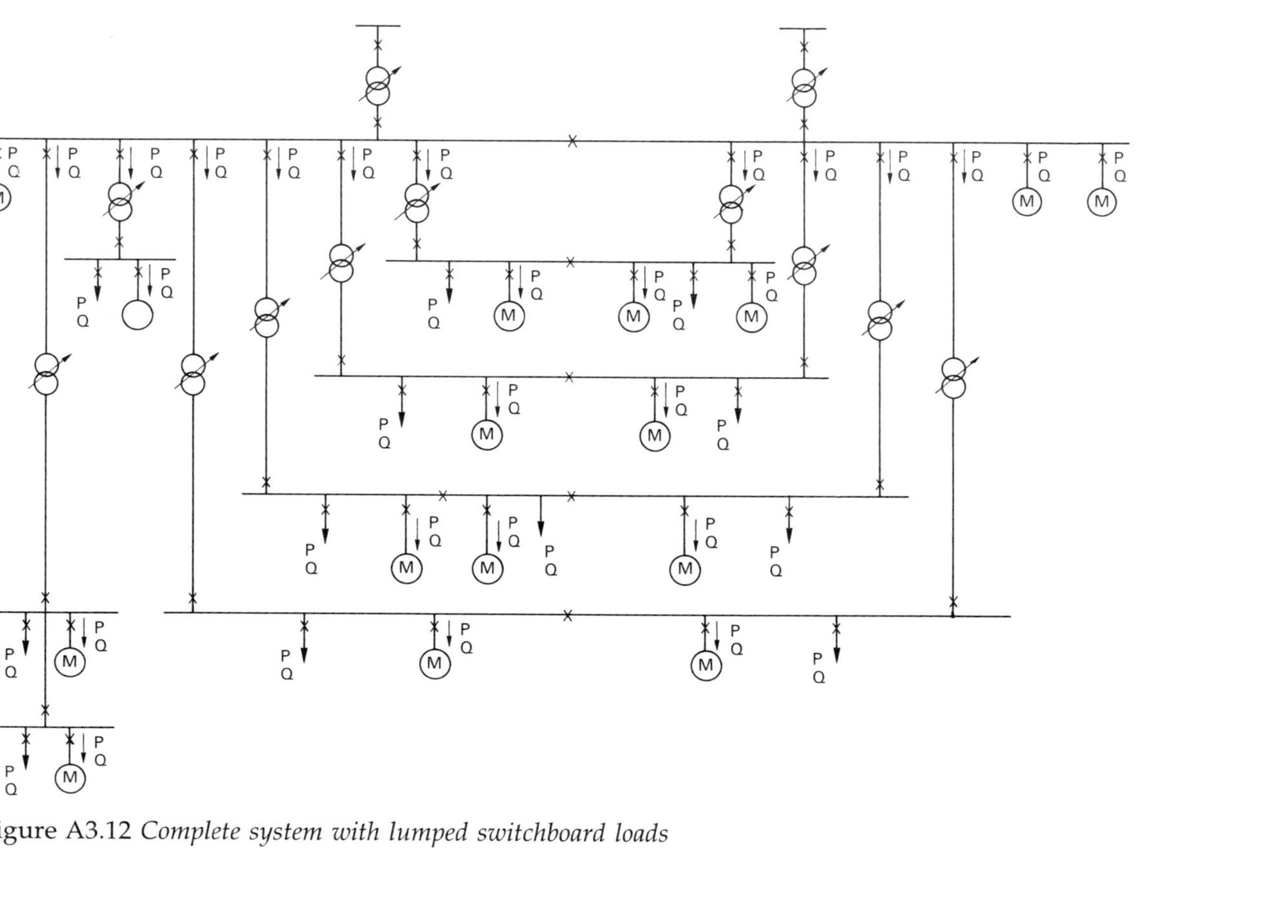

Figure A3.12 *Complete system with lumped switchboard loads*

Even small systems contain large numbers of individual units, such as generators, transformers, lines or cables, motors and other forms of specific loads. While it is theoretically possible to model each of these individually, the significance of many items is relatively unimportant in itself and it is usual to integrate such units into a single larger unit which will have the same effect on the behaviour of the rest of the system. The behaviour of one of the component items can be deduced from the behaviour of the integrated item if necessary. Alternatively, the largest item of an integrated item can be treated individually, leaving the residue to represent all the other items in the group: if the largest unit's behaviour is satisfactory, then that of the smaller ones can be assumed to be equally satisfactory (Figures A3.12 and A3.13). This

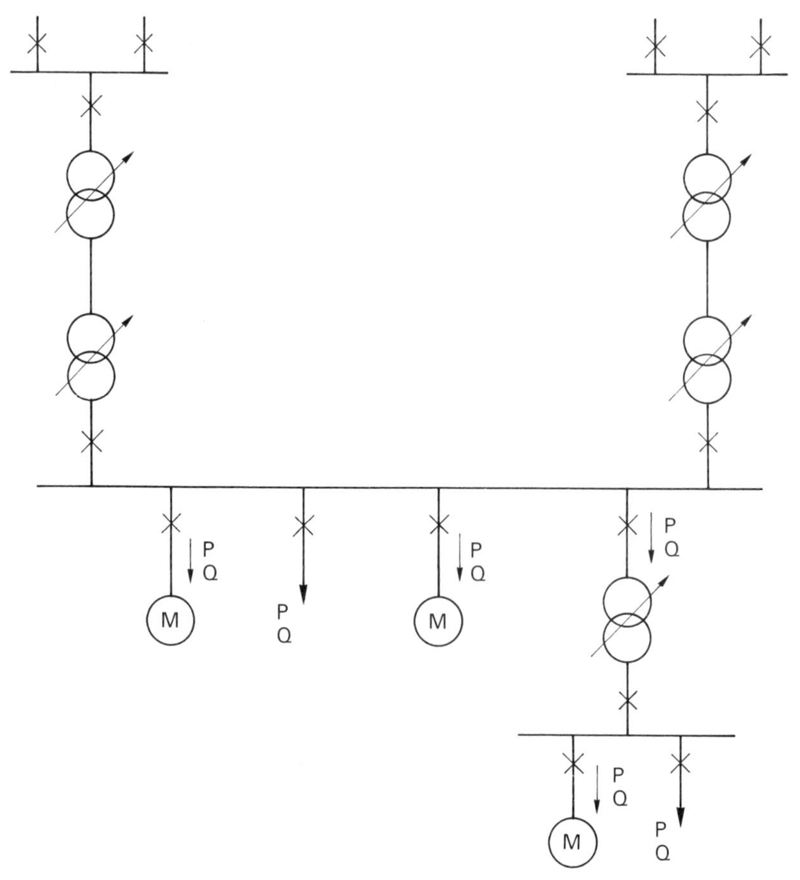

Figure A3.13 *Simplified system with interconnections to main system*

principle can be extended to all similar items of plant but the implications of doing so must always be borne in mind, particularly if the items being integrated are not identical but are only similar in behaviour. Even duplicate items do not have precisely identical characteristics, and it is important to use the most suitable values for such items, usually the most pessimistic.

Each item of equipment involved in an electrical power system study can have a large number of associated functions, many of which are interactive: for example, a gas turbine, as used for a generator prime mover, has a power output from a fuel input source and this is determined by a very complicated process involving thermal, aerodynamic and mechanical effects. It is also affected, however, by such basic factors as ambient air inlet temperature and characteristics of the fuel provided. If it is assumed that such functions are constant at the design value, the specific turbine output can be deduced from a fairly simple model; but if they require to be varied during a particular study, the simulation model required becomes very much more complicated.

Thus there are several categories of model used in such studies:

1 **Exact equipment models**. These relate input to output for one specific function or several functions, depending upon the nature of the device.
2 **Reduced function (or simplified) models**. These relate specific input to specific outputs, neglecting all other interactive functions, i.e. assuming that they are constant.
3 **Approximate models (linearized or restricted range)**. Where it is difficult to derive an accurate model representation to give an adequate prediction, it is possible to adopt a model, based on a simple relationship between the functions, which is reasonably accurate over the range of variation relevant to the particular conditions being studied, e.g. the use of saturated or unsaturated machine reactances.

In practice, each item of equipment comprises several subunits and each of these may require to be represented by a model of one of the above types. Figure A3.14 shows a simplified block diagram for a typical gas-turbine-driven AC generator set control system: the generator, main exciter and pilot exciter can be represented by suitably selected existing machine models; the automatic voltage regulator (AVR) must be represented by a control function giving a

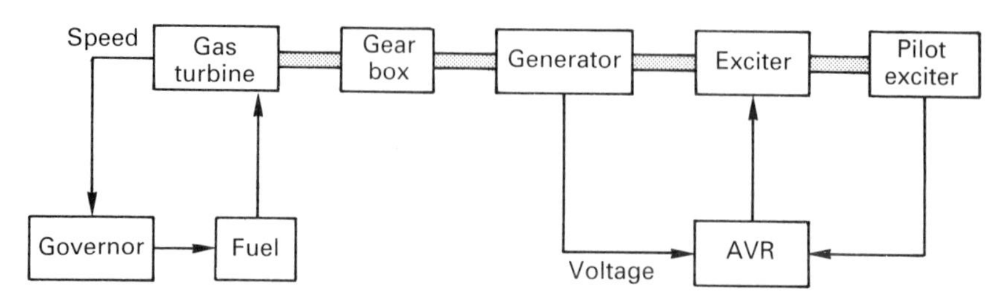

Figure A3.14 *Simple control block diagram for gas turbine generator set*

representation of the effects of generated voltage and pilot exciter outputs on the main exciter, and hence on the generator output voltage. Several excitation systems have been standardized. Figure A3.15 shows one such block diagram which uses internationally recognized symbols for the various parameters. In many instances it is found that such a simulation or model exists for the particular plant being investigated and this can result in considerable saving of time when performing studies on the plant. Similar standardized block diagrams exist for gas turbine fuel and governor systems and these usually give satisfactory results for conditions in which the main parameters do not vary significantly from the nominal operating conditions.

Additional control functions or practical limits to the various functions can be added to such simple simulations, and protective functions may also be included to cope with abnormal or fault conditions.

When the range of function variables exceeds the nominal working range, the accuracy of such simple models is reduced and may involve consideration of other control functions neglected in the interest of simplicity. Thus the behaviour of some AVRs can be distorted when the waveform of the reference voltage is severely distorted by harmonics, and gas turbines operating over wide ranges of parameters may have to have the effects of inlet guide vanes, or blow-off valves, incorporated in a satisfactorily accurate model.

Accurate simulation of each unit of a system and its function is possible but may require a very complicated model. This could involve considerable cost and time in developing and validating a suitable model to deal with all possible operating conditions. Thus, while the degree of complexity of modelling used in any study must be adequate to deal with the complete range of variables to be

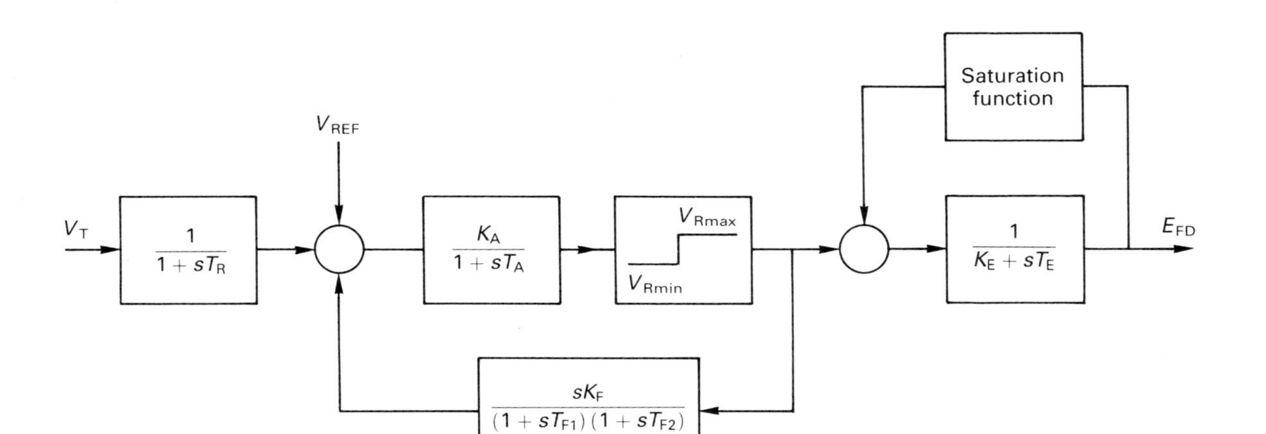

Figure A3.15 *Standardized excitation block diagram*

encountered, and must also be compatible between all main models and submodels (i.e. must possess the same order of accuracy of simulation for each unit), it is also desirable to keep models as simple as possible, when they, in turn, comprise a small part of the total system being studied.

On large systems where it is necessary to use simplified models of major units it is customary to determine the validity of selected subunits by doing 'spot-checks' to investigate their behaviour over the required range of variables determined by the main study, by simulating their function behaviour using a more accurate simulation. Such an approach can also be used to calibrate the performance from typical tests on the actual components and this can serve as a check on the overall integrity of the results obtained from the study.

In many instances, such as control circuitry, it may be difficult to obtain an accurate model valid over the complete function range. It is often acceptable to replace the model by a simulated functional relationship, i.e. to derive an equation or set of curves etc. which will give an acceptable relationship between the required input and output. These can be quite accurate over the specified range where interactive functions are negligible and can frequently be obtained by testing actual plant in operation.

When the range of possible interactive functions is extensive it may be necessary to develop an accurate model of the control device, such as an engine-governing system, or a generator excitation control system, and from this model derive simple function relationships which give an accurate representation over a required restricted range of function values.

There must always be a compromise, when modelling any system, between excessive detail of representation, which involves considerable computer data-handling and storage capacity and computation time, and oversimplification, which can result in reduced accuracy of performance prediction. Any simplifications adopted to produce an economically viable model must be matched in expected accuracy in all areas relevant to a particular study. Thus good governor simulation is essential when studying transient recovery of electrical machines, whereas the load impedance simulation need not be so accurate for this condition.

Practical operating experience with the particular condition being considered is necessary when adopting simplified modelling or simulation, and comprehensive validation is also important to ensure that the results will be accurate over the required range of variables.

One important aspect of modelling is to examine associated output function data in addition to those primarily being investigated. Thus it is necessary when studying the recovery of a large induction motor following clearance of a severe system fault not only to determine that the motor speed returns to normal but, in addition, to record the behaviour of the generator prime mover, including governing control, to ensure that their speed characteristics, which affect the system frequency, are realistic in time, since these factors have a great effect on the motor recovery, and, unless they are accurate, any prediction of motor recovery could be fallacious. This factor is of prime importance in all transient studies, where correct time prediction is often more important than function magnitude determination.

Before performing any simulation study it is important to determine exactly what information is to be obtained and also what accuracy is necessary. In many instances only one parameter is really significant; in others, it may be difficult to predetermine which are important and which need not be considered. From this the type and complexity of the necessary simulation can be determined, and this may enable a considerable part of a study to be performed using simple models, hence saving time and money. However, where this approach is not satisfactory and an extensive accurate simulation is required, it is sensible to arrange this to make available a wide range of output data covering all aspects of performance. Future study of this may bring to light some factors which had not been envisaged when the study was planned.

Figure A3.16 shows the result of a simple motor-starting problem at a remote load location from the generator G. The transformer shown has been used to elevate the receiving-end voltage above normal prior to motor starting in order to keep the value to at least 85 per cent of normal at the instant of switch-on, thus enabling the motor to generate a satisfactory break-away torque. After the motor has accelerated up to speed, the receiving-end voltage can be restored to normal. It will be noted that the existing motors M_S and M_R at the sending and receiving ends recover stability following the starting of motor M. Figure A3.17 shows the result of a short circuit fault at the sending-end busbars which reduces the M_S voltage to zero during the fault clearance time of the protection: the voltage then gradually recovers. The receiving-end voltage does not drop to

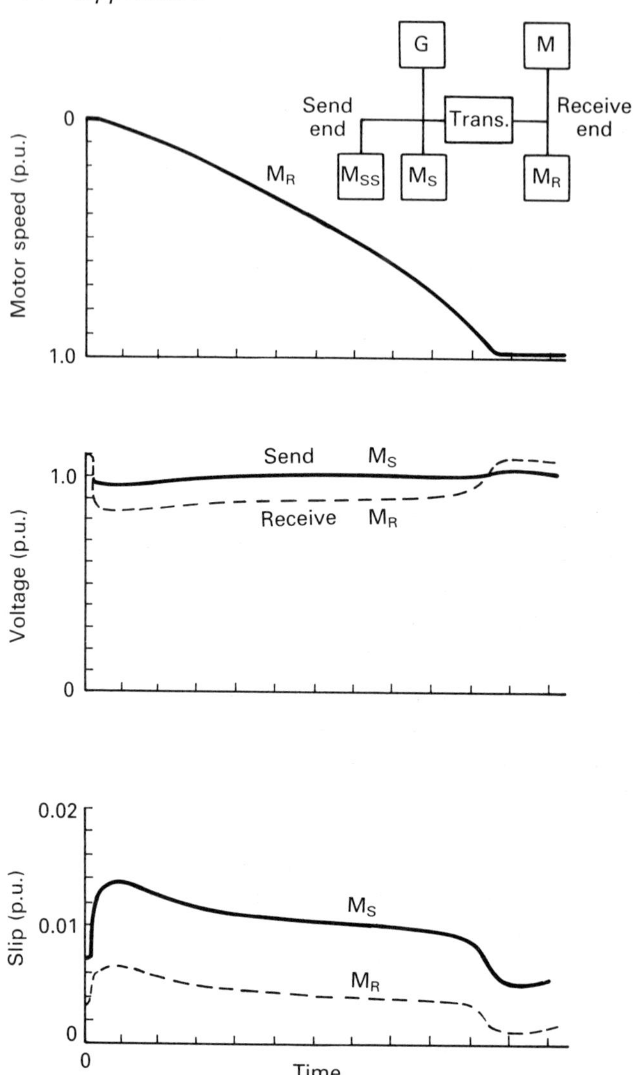

Figure A3.16 *Results of a simple motor-starting study*

zero but takes longer to recover to normal. The sending-end load M_{SS} is shed at the fault initiation time and this reduction in load enables the generation to reaccelerate the other sending-end drives M_S, with only a small slip increase, and also the receiving-end M_R, which have suffered a greater slip reduction. The two recovery curves relate to motor groups having different dynamic parameters.

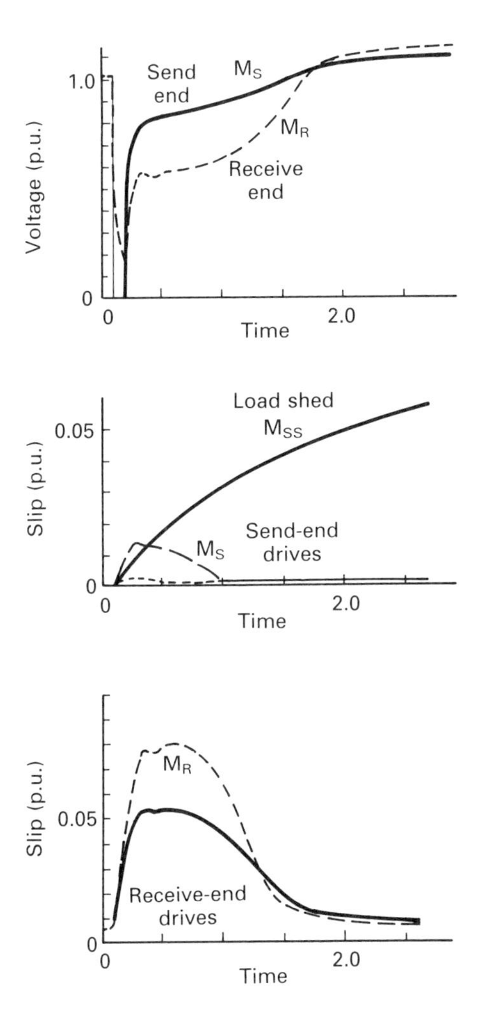

Figure A3.17 *Results of a short circuit fault at the sending-end busbars*

Figures A3.18 and A3.19 represent a slightly more complicated system intended to investigate the parameters determining stability of the receiving-end motors M following a short circuit at the intermediate generation busbar L_2. Such a study requires some additional output information to determine the performances of the two generation systems in addition to the behaviour of M. Figures A3.18(a) and (b) indicate the frequency and voltage at the three locations for the most severe Condition 1 (see Figure A3.19) which relates to the larger proportion of the total generation being at G_1 and the fixed load L_3 being a maximum. Under Conditions 1 and 2

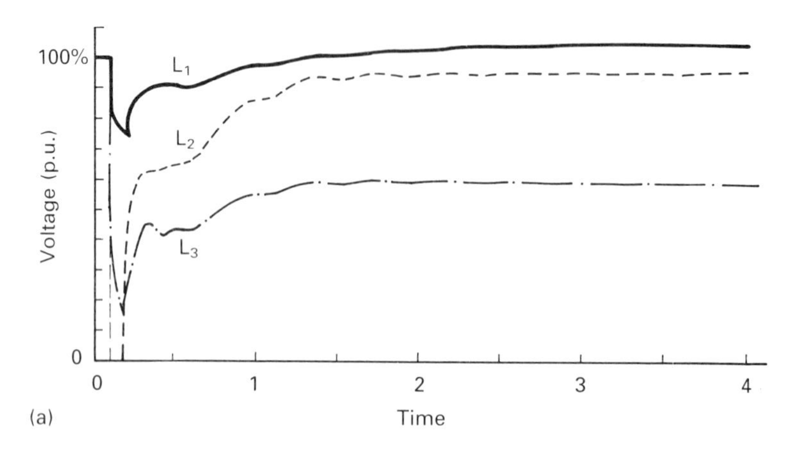

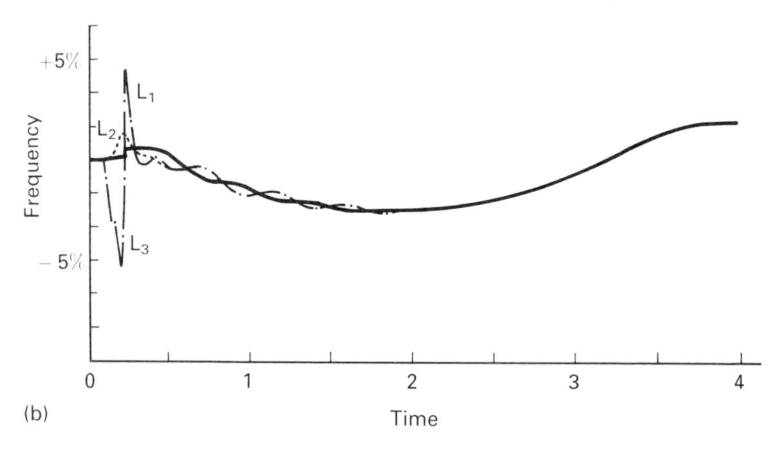

Figure A3.18 *Voltage and frequency response following short circuit*

the motor M does not recover, and recovery is progressively improved as the load L_3 is reduced. Intermediate values of generation ratio and L_3 load can be deduced, if required. Such a study is usually performed at the extreme loading parameters envisaged and also at reduced values to produce a matrix which can be used for any practical set of parameters to determine whether stability exists.

Protection systems

When performing a simulation study it is usual to exclude the function of protective devices in order that a better understanding

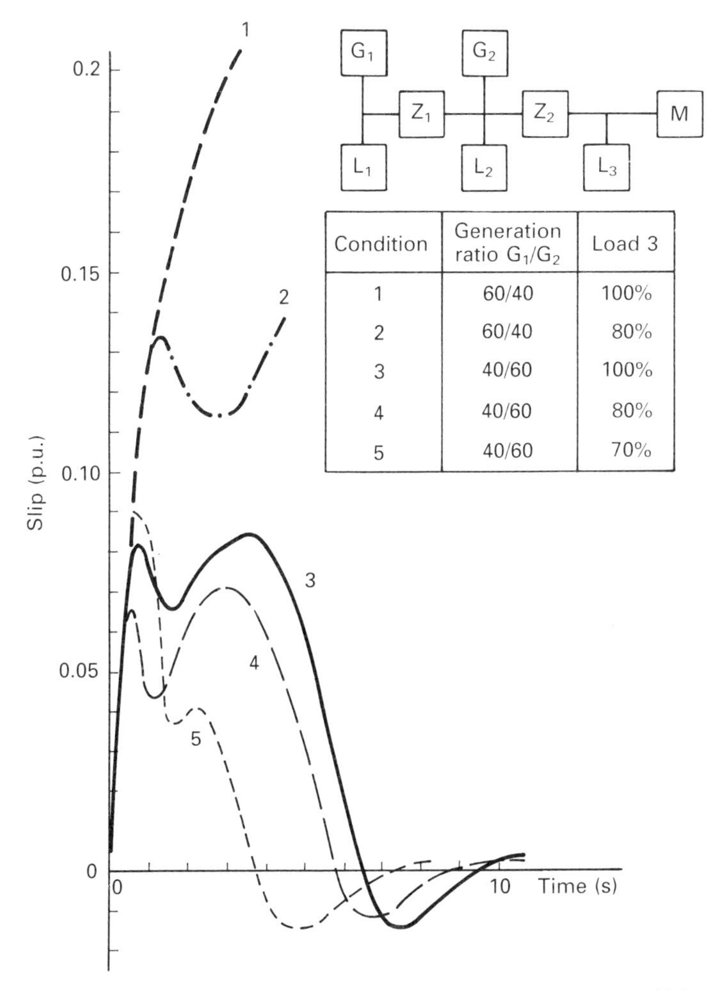

Condition	Generation ratio G_1/G_2	Load 3
1	60/40	100%
2	60/40	80%
3	40/60	100%
4	40/60	80%
5	40/60	70%

Figure A3.19 *Response of motor for alternative system conditions*

may be obtained of the prospective plant behaviour. Such study results are then used as a basis for determining what protection equipment and settings are the most suitable. However, in some types of study it may be desirable to incorporate protective features in the modelling to give a true prediction of plant capability: thus the effects of blow-off valves in gas turbine prime movers on the turbine behaviour can be included in the machine model.

On electrical systems it is usual to implement emergency actions if the system frequency drops excessively, and this type of control function can be included in the simulation. However, it is preferable to perform a study without such actions included to determine the extreme conditions encountered, as these may not be severe and the special protective function may be proven unnecessary. Should the conditions be unacceptable, however, the effect of the protective action can be determined in detail by adequate modelling incorporated into the original system.

Accurate system performance simulation is important when selecting suitable protection equipment and its setting to give maximum benefit. A typical example of this is the contribution to system fault level of induction motors. This was neglected in the past, as it was assumed to have negligible effect on switchboard ratings with regard to faults. However, as the operating times of

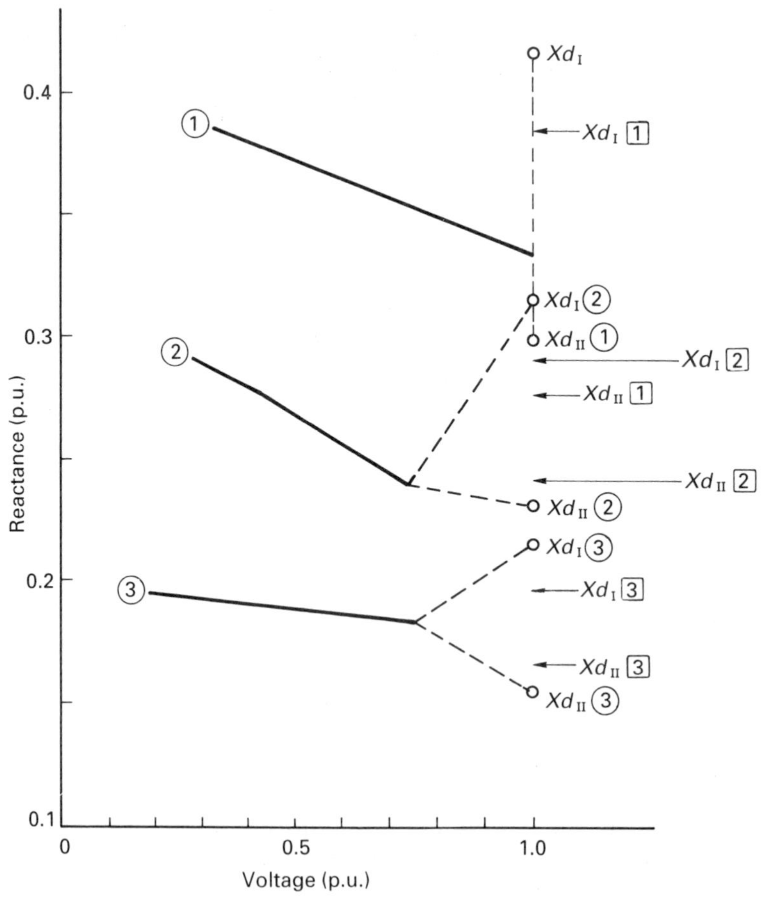

Figure A3.20 *Fault impedance of induction motors*

protection devices and circuit breakers were reduced, it was realized that this contribution could be significant. A simple model was then adopted to represent the induction motor contribution, namely its standstill impedance, which was an easily obtained value in general. As the importance of this contribution increased, a more accurate model was derived and validated from actual test results. Where the contribution to a system fault level of induction motors is likely to be important (i.e. to bring the system fault level close to the switchgear capability), then use of the more accurate simulation is justified. Figure A3.20 compares the test results and calculations for three large induction motors. The standstill impedance values usually have to be measured at reduced voltage (see curves 2 and 3). The rated voltage is deduced by extrapolation, which is difficult, as the effect of saturation can be very great; hence the deduced value of short circuit contribution can be significantly in error. The values of Xd_{I} and Xd_{II}, deduced from the motor design parameters, are shown for the three motors and the extrapolated standstill impe-

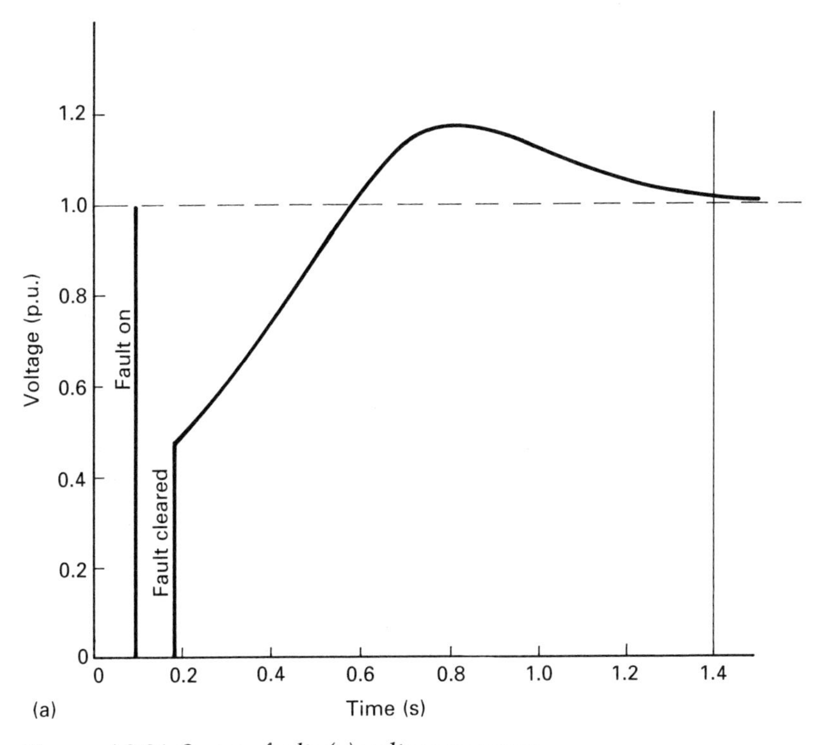

Figure A3.21 *System fault: (a) voltage response*

dances are intermediate between these two values. The arrowheads indicate the results obtained from short circuit tests on the motors. These correspond reasonably with the component values deduced from the machine parameters, which are quite consistent, whereas they do not give any consistent relationship with the extrapolated standstill impedance tests.

There is a further aspect of accurate simulation that can give all the relevant functions following any fault from a particular system condition. It enables the most suitable from of protection and also determination of the corrective actions that would be most beneficial, and their initiation. The data provided also enable optimum setting values to be determined. Figure A3.21 shows the

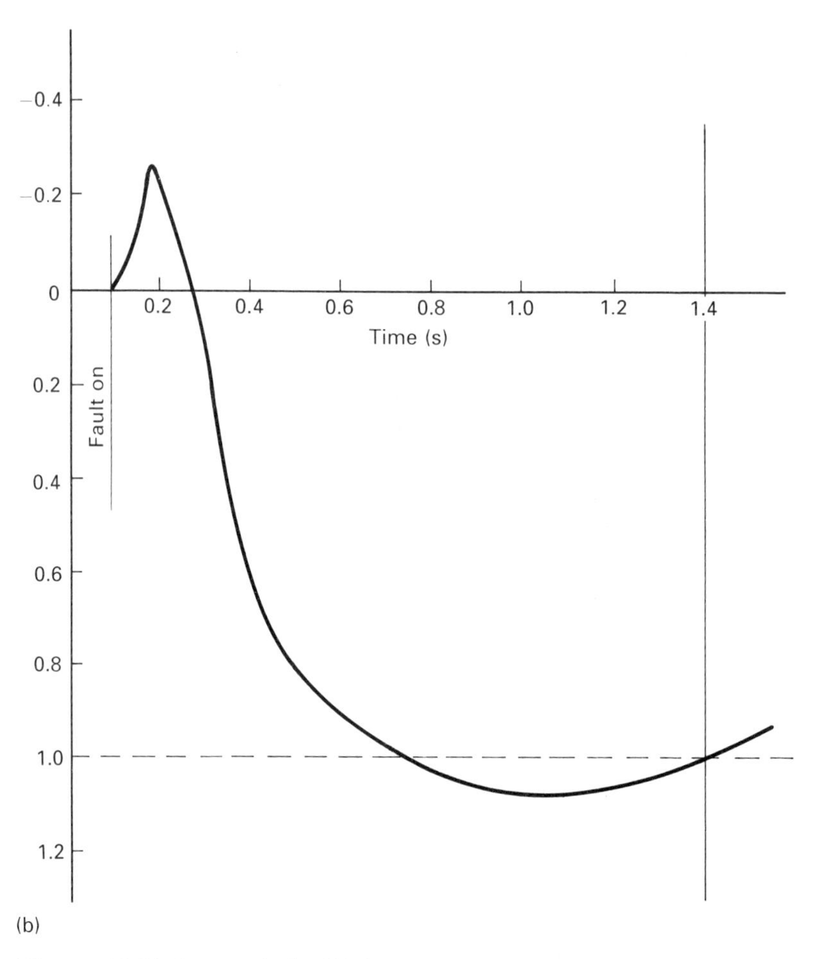

Figure A3.21 *System fault: (b) frequency response*

result of a system study involving very fast fault clearance times using bus-zone protection to ensure system stability. The system frequency rises from the fault inception at 0.1 s until it is cleared and thereafter the voltage recovers and the frequency starts to drop. The excitation system restores normal voltage in about 1.5 s and by then the governing system has limited the frequency drop to just over 1 per cent slip and started to restore it to normal. The behaviour of the various loads supplied from the system are included in the study, but unless secondary reactions were expected in the system this would not be necessary for the purpose of determining system stability as affected by generator behaviour.

Types of study

Depending upon the conditions being investigated, the extent of function interaction varies in significance. It is usual, therefore, to divide any large system study into individual aspects, where the effects of interaction can be reduced to a negligible level, or where knowledge of the significance of such effects is adequate. Thus, when considering any electrical network system, steady-state load–flow studies are first performed to ensure that the various system parameters can be kept within the plant design capabilities, to determine what values of variable factors, such as transformer ratios, are required to meet operating requirements and to establish system voltage contours. Such studies are required to establish the starting conditions for all other types of study; they can omit all transient effects, including electromechanical effects, and can use simple prime mover outputs and electrical load inputs which are known to be applicable during the conditions being studied.

Fault level studies are required to start from some specified load–flow condition but are then expected to simulate all electrical transient effects. However, the time deviations involved are so short that electromechanical effects can be ignored without significant loss of accuracy when determining fault levels. Such studies have two uses. The first is to ensure that the prospective fault level encountered by items of equipment does not exceed their capability. Great accuracy is not necessary unless the working margin is small, when maximum accuracy will be required. Thus a simplified system representation is often carried out first, to decide whether the margin between actual level and plant capability is generous or only marginal – when an accurate simulation will be performed.

The second use of such studies is to obtain specific values of functions during known fault conditions in order to determine how the appropriate protection devices will operate. For such a purpose accurate values are required, since overestimation or underestimation of values can result in faulty setting of protection.

If it is necessary to investigate transient conditions which last for such a time as to require simulation of electromechanical effects, a range of transient stability studies will be required. These usually prove to be the most difficult because such a wide range of items is involved, each with unique parameters and characteristics; these can include electrical, mechanical, and thermal effects and involve interaction between these. It may be necessary to involve different model functions for each different condition required to be investigated, should a single model not be practicable for all conditions. To keep these simulations as simple as possible, it is necessary to determine the relevant variables for each condition and select appropriate models which will give accurate results over the range of functions to be involved. It is also necessary to select those conditions that will prove to be the most serious or the most limiting, as it is impracticable to examine every possible set of conditions. As a simple example, the shaft system of a prime mover, gearbox and generator can be regarded as solid for most electrical system studies, but when it is desired to determine gearteeth loading, then accurate values of transient torque are required and the shaft system must be represented by a comprehensive electromechanical model involving all component inertias, stiffness, damping etc., many of which are nonlinear functions. To include a model of such detail in a complicated network analysis study could be quite laborious. It is often practicable to use the simple shaft model to establish accurate interfacing functions between the electromechanical system and the electrical network and apply these to the accurate mechanical model. Similar procedures can be adopted for close investigation of particular components or functions where effects are required to be considered in fine detail.

This aspect of transient analysis requires a wide knowledge and experience of similar problems and their studies to enable a representative selection of conditions to be made, together with suitable forms of models and simulation. Some conditions can be predicted as inherently unstable and it is then pointless to waste time and effort on performing unproductive studies. It is usual to select those conditions which might be unstable, as a starting point.

Should these prove unstable, it will then be necessary to start studying more stable cases to determine the limits of stability of the system. If the case, however, proves to be stable it is then possible to carry out further studies on more severe conditions to prove whether these are also stable. The same principle can be adopted when attempting to establish the limiting conditions which keep particular functions within their required operating limit.

Because of the possible complexity of modelling required, it is essential to include an accurate validation of the values and configurations adopted. This should be done by using the proposed simulation to predict a condition which has been encountered and measured on the actual plant in service or on test, or, if this is not practicable, on similar plant. This is usually a simple procedure for single items of plant which are available for testing, and frequently suitable type-test results are available for standard components, but it may be more difficult on existing installations. However, it is usually possible to perform limited tests on plant in service to validate simulation over at least part of the required operating range, thus gaining confidence in its use over the complete range. Usually an accuracy within ±2½ per cent can be obtained for a function when using good simulation.

Other types of study

The conditions considered so far have assumed that the components of the system are healthy and will respond normally to the conditions being imposed, such as starting a large motor or applying a solid short circuit to the electrical system. It is sometimes necessary to perform sensitivity studies to determine the significance of variation of one parameter for the overall response of a system, such as the ramp response of an engine governor. These are simple to perform when only a single parameter is varied, requiring only repeated runs of the same system with the necessary range of the parameter under examination. However, when the sensitivity of more than one parameter is required simultaneously, the amount of computation increases rapidly.

Other investigating studies may be required in special cases. These may require very different forms of simulation and they are usually much more complicated. Conditions which are commonly encountered, in this category, are those in which the assumption of phase symmetry and sinusoidal waveforms is not valid. For example, many system faults when initiated occur between system lines or between line and neutral. These cannot be analysed accurately using simple phase representation of system components and values of current and voltage. Such faults can be analysed in a similar manner to steady-state asymmetrical conditions by using the concept of symmetrical components for the variables and substituting appropriate component impedances for the simple symmetrical phase values.

Another instance of asymmetry arises during a machine short circuit when the voltages and fluxes associated with the different phases must be different because of their relative phase displacement with time, and this results in DC components occurring in some phases. In the extreme case one phase is fully asymmetrical, as shown in Figure A3.22. When the decrement of the DC component is longer than that of the AC component, this phase current may not pass through a current zero for several cycles (eight or nine, as shown in this instance). This feature has a significant effect on any switch or circuit breaker attempting to interrupt such a current. It is important to determine whether such a condition could arise when assessing whether a particular switch is capable of performing such fault interruption.

When determining the breaking duty it is necessary to establish the total AC and DC components of current occurring at the location being considered. This requires not only an accurate magnitude–time relationship of all component currents but also their exact phase relationship to each other. Accurate data are needed with regard to both the resistance and the reactance of all circuits in the electrical network and also of all the component items of plant. As a consequence, the amount of data required for such a study, and the computation involved, is much greater than that required for a simple symmetrical assessment. For this reason, such studies are only performed when there is a reason to believe that breaking duty could be significant. The DC time constant is affected by the resistance in the network. In most systems the distributed circuit resistance is usually high in relation to the machine resistance; this results in a reasonably short DC time constant which usually precludes non-zero fault currents. However, in small compact systems, such as offshore oil production platforms, where the distribution resistance loss is of the same order as that of the machines, the condition can arise where the significant number of

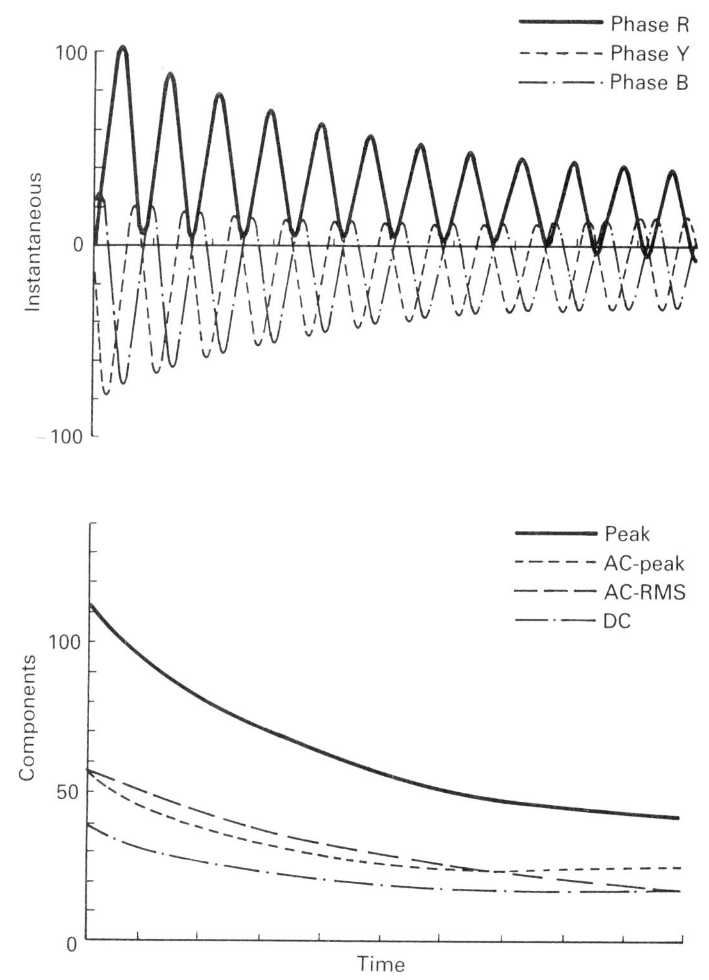

Figure A3.22 *Asymetrical fault current*

non-zero current cycles can prove limiting on the type of switchgear which would be suitable for such an application.

Even when the idealized conditions of constant frequency and purely sinusoidal voltages and currents are not encountered, most normal simulations remain valid for all practical purposes. However, conditions can occur when deviations from the ideal become significant and require special investigation when studying specific functions.

One of the most common of these is the presence of one or more specific harmonics in the system currents and voltages. In the past some harmonics were found to interfere with communications systems, and telephone interference factors (TIF) were established to determine acceptable levels of the relevant harmonics; it was then necessary to establish the harmonic spectrum for the system. In recent years, with increasing numbers and ratings of static converters, other adverse effects of harmonics have become much more significant. Electricity supply companies now impose strict limits on the harmonic spectra content of loads connected to their system. However, when considering small generation systems these same limits are not necessarily relevant. If implemented, they might be unnecessarily restrictive in some installations, whereas in others there may be unique items of plant which are peculiarly susceptible to particular harmonics and which might require the limitation of specific harmonics to lower values.

All such systems require some form of harmonic distribution study and this requires detailed data on the sources of all harmonics, the impedances of all devices and elements at each relevant frequency, and also the susceptibility limits of any special items of plant such as filters, protection or monitoring equipment which could be damaged or have their operation impaired in any way.

When it is only necessary to investigate a few specific harmonics, this can be done by treating them individually using the fundamental system representation with parameters adjusted appropriately, and then summing the total effect. Such a procedure

Table A3.3 Typical generator harmonic impedance spectra

Harmonic	Impedance p.u.	
	d Axis	*Q* Axis
5	0.26 + *j* 1.09	0.80 + *j* 0.93
7	0.36 + *j* 1.53	1.11 + *j* 1.31
11	0.49 + *j* 2.15	1.08 + *j* 2.00
13	0.58 + *j* 2.54	1.27 + *j* 2.35
17	0.68 + *j* 3.13	1.27 + *j* 2.94
19	0.76 + *j* 3.50	1.42 + *j* 3.29
23	0.85 + *j* 4.07	1.42 + *j* 3.86
25	0.92 + *j* 4.42	1.54 + *j* 4.20

involves many assumptions and leads to inaccurate results, but such studies can often indicate whether a serious problem exists or not, and then a more accurate study can be implemented if necessary. An accurate study is very expensive and time-consuming, as each item of data must be of equal accuracy and generalizing assumptions cannot be adopted. For example, it is inadequate to assume that all reactances are proportional to frequency (see Table A3.3, giving typical generator data); it is usually necessary to obtain accurate data from manufacturers' design or test data. Other devices have no single set of parameters: thus a converter-fed drilling motor will produce a different harmonic spectrum for different values of torque and speed and in service this will vary during each drilling cycle. It is therefore necessary to determine a characteristic spectrum giving the worst effects on the system. Where more than one such drive is fed from the same supply, the natural diversity of loads gives a measure of averaging. Figure A3.23 indicates a range of possible spectra for different converter/drive units, and they can be seen to follow a common trend which can reasonably be generalized, ensuring that the results of the harmonic study will be pessimistic.

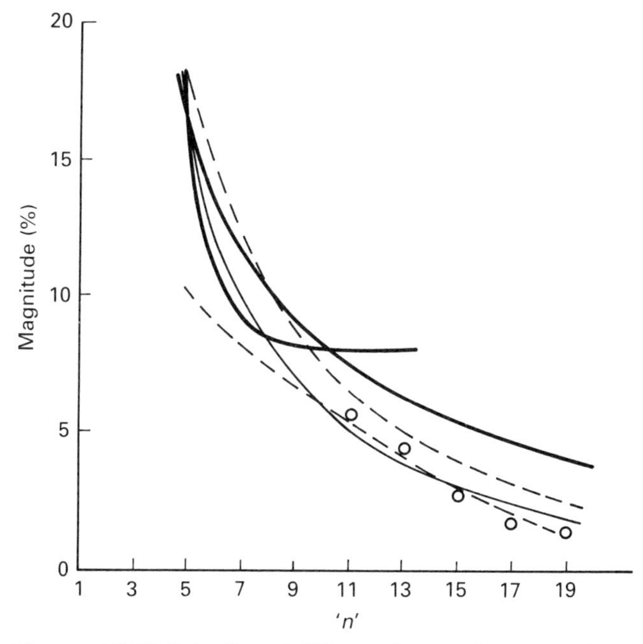

Figure A3.23 *Selection of different harmonic spectra for converter drives*

Where accurate results are required, the simulation must identify not only the harmonic magnitudes but also their relative phase relationship. This procedure can be used to examine pulse-type functions, e.g. arcing faults or voltage surges such as those which can be produced by converters. Figures A3.24 and A3.25 show typical results from such studies.

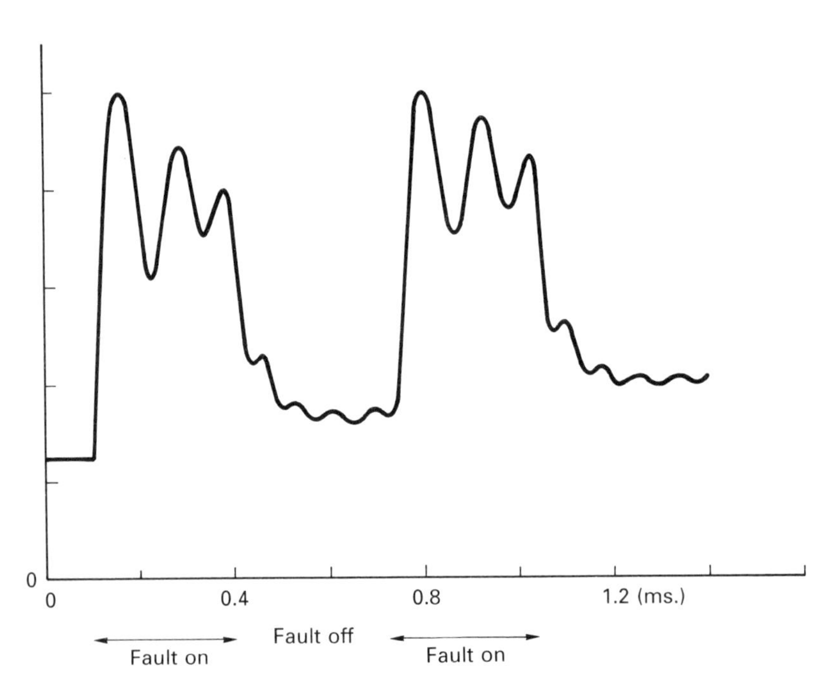

Figure A3.24 *Converter fault waveform*

Another set of conditions is involved when frequency changes occur in the system. Transient disturbances produce a frequency-change pulsation which usually disappears within a few seconds, and is affected significantly by the electromechanical characteristics of the components of the system. However, there are conditions under which sustained frequency pulsations can occur. These may be excited by a prime mover, such as a reciprocating engine which produces not a constant torque during each engine cycle, but a mean torque plus a superimposed pulsating torque at a frequency determined by the engine speed, number and distribution of cylinders and number of strokes. Such a harmonic torque will result

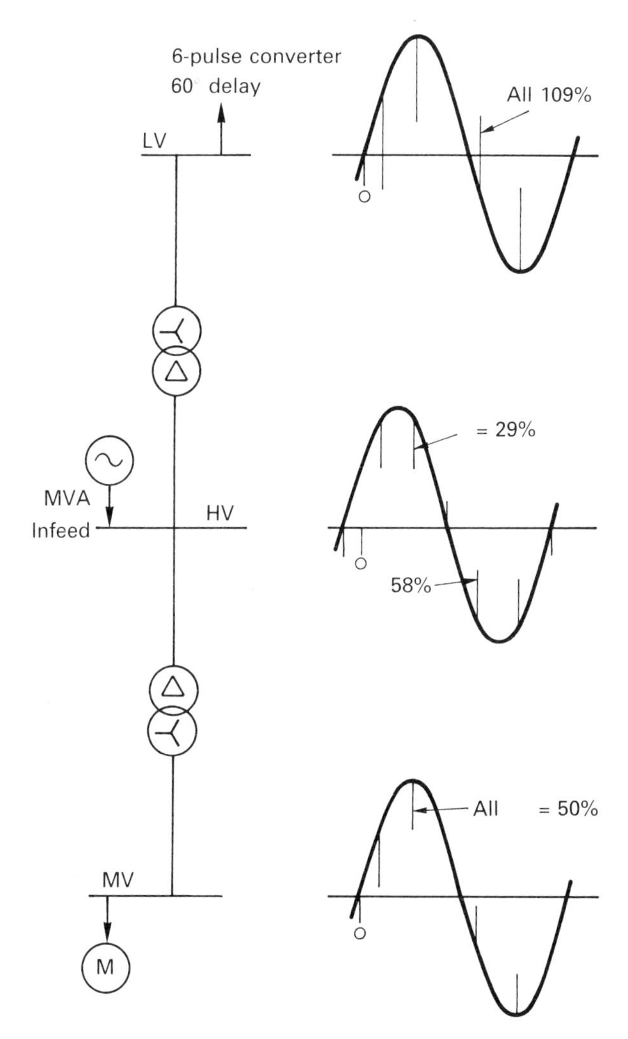

Figure A3.25 *Simulated voltage notch analysis*

in cyclic irregularity in the generator rotor which will in turn produce a voltage fluctuation and also a power output fluctuation. In many situations these pulsations are negligible and may not even be noticed by the casual observer. However, they can behave as an exciting force, and any component in the total system which is susceptible to resonance at this frequency can be excited. If sufficient damping is not present, the magnitude of the resonance can produce dangerous conditions which can easily result in plant failure.

The incorporation of additional inertia in the prime mover shaft system can reduce the angular pulsation produced by the same value of pulsating torque, but there are practical limits to how much can be added to any particular set. Should further attenuation be required to effects produced on the electrical system, it is possible to use the generator excitation system to modulate the voltage generated in tune with the angular pulsation so as to further reduce the consequent voltage magnitude of pulsation (although this does not reduce the angular pulsation).

A similar phenomenon occurs on electric motors driving reciprocating compressors that absorb an irregular torque cycle from the motor, which in turn demands an irregular current from the supply. This current, flowing through any impedance such as a transformer, produces a voltage fluctuation and this can have similar effects to those produced by a generator with a reciprocating prime mover. It often results in serious fluctuation in the level of illumination produced by some forms of lighting. A simple compressor factor can be determined for any motor, depending upon its electrical 'stiffness' and its mechanical inertia, and this gives the relationship between the worst harmonic torque spectra required by a specific compressor and the motor current pulsation drawn from the supply. Standard compressors have known limiting torque spectra and can identify the 'compressor factor' required by a suitable driving to obtain an acceptable value of current pulsation. Standard pulsations of ±66 per cent, ±33 per cent and ±20 per cent are found to be acceptable for particular motor locations in the electrical distribution system in relation to loads particularly susceptible to voltage fluctuation. Such an investigation simplifies the selection of a suitable motor for a particular compressor when supplied from an appropriate system configuration. When such a simple simulation is inadequate, either because adequate data are not available or because the sensitivity of plant in the system is particularly high, an accurate simulation must be performed. This requires the complete set of compressor torque spectra for each required operating condition, together with an accurate electrical simulation for the motor and the adjacent supply system. It also requires an adequate simulation of the mechanical shaft system, including the compressor, any couplings or gears and the electric motor. When only electrical effects need investigation, the mechanical system can be represented as a block unit system; but when mechanical torques require analysis, e.g. if shaft torsional problems are suspected, a detailed simulation of the total shaft system will be required to identify each discontinuity uniquely (e.g. each cylinder in the compressor) so that their loadings can be incorporated individually.

Such shaft systems which incorporate electrical machines may require investigation even when the consequent effects on the electrical system are not significant, because torsional resonances may occur due to periodic or pulse-exciting torques generated either by the electrical or mechanical machines. Other mechanical components forming part of the electromechanical set may also be excited by such torsional irregularities and, if they are resonant at the particular frequency, they can fail: for example, fan blades mounted on a generator rotor can suffer fatigue failure from such phenomena.

It will be appreciated that where accurate simulations are being used on systems in which several electrical machines are connected to a comprehensive electrical system simulation, the amount of data to be handled and the extent of computation can be enormous. However, where the electrical units have different electromechanical parameters and consequently different resonance frequency spectra, it is necessary to use such highly accurate data. It is usually preferable to employ simplified simulation in the main unit, to try to determine whether any prospective dangerous modes are likely to occur, and then to take steps to modify the parameters either at the design stage, or by modifying control procedures to eliminate the possibility of operation in the dangerous mode. This procedure is one of the most important when designing components for any electrical system either at the initial design stage, or when adding plant to such a system. Thus by careful system design it is often possible to reduce the extent of simulation required for any system study. In many instances it can be avoided altogether, but only after full consideration has been given to all possible failures or dangerous modes that could be expected to arise at any time in the future and should not be done by neglecting to consider all the facts relevant to the system and its component items of plant.

Index